Praise for *Beyond Mea...*

"Fascinating. . . . *Beyond Measure* offers engrossing accounts of the role that measurement has played in scientific progress. . . . [James] Vincent also presents a deep history of measurement's role in society."
—Christie Aschwanden, *Undark*

"Fascinating. . . . [A]s an account of the lengths humanity has gone to in the name of measurement, this quirky history is inch-perfect."
—Chris Allnutt, *Financial Times* (UK)

"Vincent has an eye for fascinating facts, but also a deeper purpose: the story he tells is one in which control has always been bound up with the drive to measure and to collect data."
—Richard Dunn, *Times Literary Supplement* (UK)

"Superb. . . . Telling the story of metrology is not easy [but] Vincent is equal to the task."
—Tom Whipple, *Times* (UK)

"A pacy romp through time and space . . . *Beyond Measure* documents humanity's attempts to claw dependable truths from a chaotic universe."
—Chris Stokel-Walker, *New Scientist*

"A hugely ambitious work encompassing a vast sweep of scientific progress and human endeavour. . . . [I]mpressive, enjoyable and immensely thought-provoking."
—Susan Flockhart, *Herald* (UK)

"A remarkable story of human endeavour, experiment, and belief. . . . [A]n erudite and elegant read."
—Simon Humphreys, *Mail on Sunday* (UK)

JAMES VINCENT

BEYOND MEASURE

The Hidden History
of Measurement from Cubits
to Quantum Constants

W. W. NORTON & COMPANY
Celebrating a Century of Independent Publishing

For information about permission to reproduce selections from this book, write to
Permissions, W. W. Norton & Company, Inc., 500 Fifth Avenue, New York, NY 10110

For information about special discounts for bulk purchases, please contact
W. W. Norton Special Sales at specialsales@wwnorton.com or 800-233-4830

Manufacturing by Lakeside Book Company.

Library of Congress Cataloging-in-Publication Data

Names: Vincent, James (Journalist), author.
Title: Beyond measure : the hidden history of measurement from cubits to quantum
constants / James Vincent.
Description: First American edition. | New York, NY : W. W. Norton & Company, 2023. |
"First published in the UK by Faber & Faber Ltd. under the title Beyond Measure: The Hidden
History of Measurement"—title page verso. | Includes bibliographical references and index.
Identifiers: LCCN 2022027408 | ISBN 9781324035855 (cloth) | ISBN 9781324035862 (epub)
Subjects: LCSH: Weights and measures—History. | Measurement—History.
Classification: LCC QC83 .V56 2023 | DDC 530.8/109—dc23/eng20221007
LC record available at https://lccn.loc.gov/2022027408

ISBN 978-1-324-06614-9 pbk.

W. W. Norton & Company, Inc., 500 Fifth Avenue, New York, N.Y. 10110
www.wwnorton.com

W. W. Norton & Company Ltd., 15 Carlisle Street, London W1D 3BS

1 2 3 4 5 6 7 8 9 0

For all my teachers

CONTENTS

CONTENTS

INTRODUCTION

Why measurement matters

The very first measurement, like the first word or first melody, is lost to time: impossible to localise and difficult even to imagine. Yet it was a hugely significant act: another addition to that nest of primeval consciousness that grew in the brains of our ancestors hundreds of thousands of years ago, and that would eventually set us apart from the other animals of the plain. For measurement, like speech and play, is a cornerstone of cognition. It encourages us to pay attention to the boundaries of the world, to notice where the line ends and the scales tip. It requires that we compare one portion of reality to another and describe the differences, creating a scaffold for knowledge. Measurement is the root of all tectonic arts, enabling construction and urban living, and the beginning of quantitative science. If we could not measure, then we could not observe the world around us; could not experiment and learn. Measurement allows us to record the past and by doing so uncover patterns that help predict the future. And finally, it is a tool of social cohesion and control, letting us coordinate individual effort into something greater than the sum of its parts. Measurement has not only made the world we live in, it has made us too.

I first began to realise the importance of measurement when writing about the redefinition of the kilogram as a journalist in 2018. I had travelled to Paris for the assignment, and there interviewed scientists who had been working on the project for decades as part of the International Bureau of Weights and Measures, the organisation that oversees the metric system. They explained how, since the eighteenth century, the kilogram had been defined as the weight of a particular lump of metal: an actual physical artefact, kept under lock and key in an underground vault in France. Every weight in the world (even the non-metric ones) could be traced back to this single standard, to *the* kilogram, or *Le Grand K* as it was known to its keepers. Advances in technology, though, meant the kilogram no longer met society's demand for precision, and so scientists had committed to redefining its value using fundamental constants of nature, derived not from base matter but from quantum properties buried deep in the foundations of reality. What was more, they had already carried out the same substitution for every other metric unit in existence. Length, temperature, time and more – all had been silently redefined by this international conspiracy of measurement.

The existence of this hidden world was a revelation. I felt like I had opened the door of my flat one morning only to step out on to the surface of an alien planet, surrounded suddenly by strange trees and the cries of unfamiliar animals. The idea that something as fundamental and commonplace as a unit of measurement was even capable of change was thrilling, and the more I learned, the more questions I had. Why is a kilogram a kilogram, anyway? Why an inch an inch? Who first decided these values and who maintains them now?

As I followed these breadcrumbs, I began to understand what an intellectual feast measurement truly is, what a banquet of historical, scientific, and sociological wonder. The roots of measurement are entangled with those of civilisation, traceable back to the ancient Egyptians and Babylonians. It was these societies that first learned to apply consistent units in construction, trade, and astronomy, building towering monuments to gods and kings, and mapping the stars with their newfound power. As units of measurement developed they became a tool of authority, claimed as the prerogative of the powerful, who used measurement to organise the world to their liking. Similarly, the science of creating accurate measurements – metrology – is entwined with some of the greatest breakthroughs in our understanding of the natural world, and has helped to redefine our place in the cosmos many times over. More than this, measurement is a mirror to society itself; it is a form of attention that reveals what we value in the world. To measure is to choose; to focus your attention on a single attribute and exclude all others. The word 'precision' itself comes from the Latin *praecisio*, meaning 'to cut off', and so, by examining how and where measurement is applied, we can investigate our own impulses and desires.

As it stands today, the world around us is the product of countless acts of measurement, their presence rendered invisible by their ubiquity. Whether you are reading these words on the page or on a screen, their finished form is the product of careful weighing and counting. The pulp that forms the paper was made using a chemical mix finely calibrated to tease apart the wood's fibrous cells without destroying their structure. The resulting sheets were forced through gigantic metal rollers of staggering precision, squeezed to the consistent thickness you now feel between your fingers. They were cut and bound to a familiar size, before being packed, weighed, and shipped around the world. Even the font used to render these words is the product of careful measurement; every serif pruned, the gaps between each pairing of letters nudged into equilibrium. And if you are reading this in a digital format, then this chain of measure is even more complex, starting with the atomic-scale engineering of silicon chips and the carefully balanced alchemy of your device's battery. Regardless of whether we think about it or not, measurement is suffused throughout the world; an ordering principle that affects not only what we see and touch, but also the often intangible guidelines of society, from clocks and calendars to the rewards and punishments of work.

————

Measurement is not an intrinsic feature of the world but a practice invented and imposed by humanity. The earliest evidence for what we might describe as measurement comes in the form of animal bones carved with notches. These metrological relics include the Ishango Bone, a baboon fibula between 18,000 and 20,000 years old, and the Wolf Bone, older still at roughly 33,000 years in age.[1] Reading their meaning is like any augury, indefinite and intuitive, but archaeologists think the ordering of marks on these bones might make them tally sticks: the first formal measuring tools.

In the case of the Wolf Bone, its incisions are divided into groups of five, a common boundary in many numeral systems. Cultures from around the world tend to count by marking one, two, three, four, and then striking, slashing, or hooking a line for five. Psychological studies suggest this is something close to an innate cognitive limit – a natural partition in human thought, though one that is far from impermeable. When tested on our ability to count at a glance, humans can usually take in three or four items at most.[2] More than that and we need to start consciously numbering. We need to measure. These notched bones, then, may mark the moment, repeated many times around the world, that our species' ambition exceeded the capacity of our brains and we reached for external support. They show when we began to measure the world around us and, as a result, understand it better.

Knowing what phenomena were recorded by these artefacts would help us decipher the place of measurement in our early cognitive development, but without written records we can only speculate as to their purpose. Perhaps the tally on the Wolf Bone was kept by a hunter who wanted to strengthen the connection with their prey by using its bones to count their kills. Perhaps they were tracking the passage of time, with each mark representing a single day. If so, then the total number of notches – fifty-five – is near enough double a lunar month, a unit that existed in the movements of our solar system before humans ever named it. If this is the case, then it means the bones might have recorded a sacred activity rather than a profane one, as such cosmological measures are entwined with ideas of the divine and spiritual in ancient societies. Monitoring the coming and going of the seasons was a way for early humans to engage with the life-giving rhythms of the natural world, as well as a first step to taking control over them. The first calendar humans ever established would have been the calendar of the seasons: marking the passage of time via animal migration and the appearance of certain flowers and crops.

It seems quite certain that humans are the only creatures to develop a formal system of measurement. We know that many species, from rats to raccoons, have an understanding of quantity and are able to distinguish between bigger and smaller piles of food, for example, while other animals carry out feats that surely require some intuitive form of reckoning (think about the astounding journeys made by birds across continents, orientating themselves using methods not fully understood by science). But these skills are limited in scope, and indeed, studies of children suggest that measurement, like writing and counting, is a cultural skill rather than an intuition we are born with.[3]

In one study from 1960, children between the ages of four and ten were shown an 80-centimetre tower of blocks on a table and asked to build a tower of identical size a little distance away.[4] While they were working on the task, a screen was placed between their tower and the original, stopping them directly comparing the two. To solve this puzzle, the youngest children, no older than four or five years old, eyeballed the job. They looked at the first tower, then started building their own approximate copy. The next cohort, children up to the age of seven, decided that visual comparison was not enough and used their bodies as measuring sticks. They held up their arms, hands, and fingers to the two towers, and compared their heights against their own. (At this age, some children also decided, quite understandably, that the game was rigged and simply ignored the instructions and built their tower next to the original.) The oldest group, children aged seven and over, turned to external measures, using strips of paper and sticks that had been provided as makeshift rulers. And even among these there were subtle differences. Younger children were happiest using paper and sticks the same height as the towers, while older individuals were comfortable using smaller items that could be counted as subdivisions of the whole.

Studies like this suggest not only that measurement is a skill we acquire with age, but that a key component of this practice is our ability to abstract. It's not enough to simply compare one tower to another or use a measuring tool the same height as the target. We must instead create an intermediary: a unit of measure that represents nothing but its own value and provides a convenient medium for transferring information from one domain to another.

There's some suggestion that our ability to process number in this way is part of a larger cognitive trade-off made long ago in our evolutionary past. Evidence comes from our closest genetic relative, the chimpanzee, who displays a remarkable facility with certain sorts of number tasks. With the right training, a chimp can be shown the numbers 1 to 10 on a screen for just a fraction of a second before tapping the now hidden digits in the correct order, doing so much faster and more accurately than humans. In fact, chimps can complete this task even if the numbers only appear for as little as 210 milliseconds.[5]

This is less time than it would take your eye to move around the screen, which suggests the skill being called upon is not a grasp of number as we understand it, but eidetic memory: the ability to retain complex visual information after brief exposure. It's a fantastical ability, the sort we might associate with savants, but it has its own limitations. The same chimps that can perform these feats are unable to replicate other basic numeracy skills, like matching groups of items larger than four or five with the correct number, even after years of training.[6]

Researchers behind this work theorise that some common ancestor of both chimps and humans possessed an eidetic memory, which would have been just the thing for identifying threats in a jungle environment. In a flash you could take in a tangle of leaves, vines, roots, bark, flowers, fruits, and *teeth*, identifying a potential predator and raising the alarm. At some point, though, evolutionary forces nudged a group of our ancestors into trading their souped-up memory for other cognitive aptitudes, including, we think, the ability to process language; to socialise and learn from one another. These were the cognitive tools that would allow measurement to flourish, helping to construct the systems that now sustain so much of modern life.

————

It was the nineteenth-century British physicist William Thomson, better known as Lord Kelvin, who offered one of the most memorable summaries of the contribution of measurement to human knowledge. 'When you can measure what you are speaking about, and express it in numbers, you know something about it,' said Thomson, 'but when you cannot express it in numbers, your knowledge is of a meagre and unsatisfactory kind; it may be the beginning of knowledge, but you have scarcely, in your thoughts, advanced to the stage of *Science*.'[7]

Thomson's words epitomise a sort of metrological triumphalism: a confidence in the power of number to square the untidy mysteries of the universe and tame the unknown through calculation. It is a reasonable belief given the history of the sciences, where accurate measure has time and time again proved itself a prerequisite for experiment and a spur to discovery. Thomson's own groundbreaking work in thermodynamics and electromagnetism relied heavily on just such accurate observations, but this connection can be traced back much further to the discipline of ancient astronomy, a profession that blends what we would now categorise as mysticism and empiricism.

In Babylonia, a Mesopotamian kingdom that emerged around 1894 BC, the gods were believed to be regular, if oblique, communicators, bestowing everything from animal entrails to the 'colour of a dog that urinates on a man' with divine import.[8] But the heavens were thought to provide a particularly authoritative format, with the ubiquity and clarity of the stars and planets constituting a celestial PA system that broadcast far and wide. Changes in the night sky might warn of impending disaster: of disease, flood, or invasion. Or they might herald a time of peace, the birth of a long wished-for child, or the start of a profitable trading arrangement. To decode these messages, Babylonian astronomers created detailed records of celestial movements in the heavens, using the resulting tables to sieve out irregularities and, with them, the favour of the gods. This was the kernel of what we now call the scientific method – a demonstration that accurate observations of the world could be used to forecast its future.[9]

The importance of measurement in this sort of cosmic comprehension did not develop smoothly over the centuries. Indeed, in the Middle Ages in Europe, reckoning with hand and eye was sometimes seen as producing a rather shabby sort of knowledge, inferior to that of abstract thought. This suspicion was due to the influence of ancient Greeks in the era's scholasticism, particularly Plato and Aristotle, who stressed that the material world was one of unceasing change and instability, and that reality was best understood by reference to immaterial qualities, be they Platonic forms or Aristotelian causes. It would take the revelations of the scientific revolution to fully displace these instincts, with observations of the night sky once again proving decisive.

Consider, for example, the unlikely patron saint of patient measurement that is the sixteenth-century Danish nobleman Tycho Brahe. By most accounts Brahe was an eccentric, possessed of a huge fortune (his uncle Jørgen Brahe was one of the wealthiest men in the country), a metal nose (he lost the original in a duel), and a pet elk (which allegedly died after drinking too much beer and falling down the stairs of one of his castles).[10] After witnessing the appearance of a new star in the night sky in 1572, one of the handful of supernovae ever seen in our galaxy, Brahe devoted himself to astronomy. Ancient wisdom and religious doctrine held that the heavens were immutable, but Brahe castigated such 'blind watchers of the sky' – the new star plainly proved otherwise, and he spent decades compiling detailed and precise astronomical records in a purpose-built observatory named Uraniborg. It was the data collected here that would allow Brahe's apprentice, the visionary German astronomer Johannes Kepler, to derive the first mathematical laws of astronomy, the three laws of planetary motion, which correctly described the elliptical orbits of the planets and treated the occupants of our solar system as ordinary matter rather than divine or ethereal substance. It was measurement, then, that focused this new sort of attention on the cosmos, uncovering eternal truths that would eventually displace the verities of the Church.

By the time we reach Lord Kelvin, the power of measurement has proven its supremacy not just through scientific knowledge but also its industrial application. In the nineteenth century, it was precision engineering that transformed the steam engine from a leaky and inefficient machine to the high-pressure muscle of the Industrial Revolution, while the ability to accurately measure and meter electricity allowed for its commercial application in lighting, communication, and more. This was the century in which the factory replaced the farm as the backbone of a nation's wealth; in which telegraph lines connected continents and X-rays illuminated the interior of the human body. The 1800s may have begun under the flicker of torch and gaslight, but they ended with a blaze of electricity – all advances that owed at least something to measurement, and would only accelerate in the years that followed.

———

The calculations of Kepler are sometimes thought of as the first 'natural laws' of science, in that they are unchanging, precise, and verifiable. Their authority is a product of their universality: their predictions are applicable not just to *this* planet at *this* time but to all planets across time and space. In other words, they are generalised, abstract rules – qualities essential to the development of measurement. Indeed, if you were to summarise the history of measurement in a single sentence, it would be as a history of increasing abstraction. Measurement begins life rooted in the particulars of human experience but over time has become increasingly detached from our life and labour. Just as with Kepler's laws, the result is that it has attained authority over an ever-expanding domain.

As the children who measured towers in that 1960 study demonstrated, the first tools of measurement we turn to are our own bodies. Common units like the hand and foot are still in use today, while others are at least familiar, like the cubit (the length from the elbow to the fingertip) and fathom (the span of your outstretched arms). Some cultures have created particularly rich indexes of measure from the human body. The Aztecs not only had equivalents of the cubit and fathom, but also units based on the forearm alone (the *omitl*), the tip of the hand to the armpit (the *ciacatl*), and the tip of the fingers to the shoulder (the *ahcalli*).[11] The Māori have derived at least twelve units from the body, from the smallest, the *konui*, equal to the first joint of the thumb, to the largest, the *takoto*, which measures the full length of the body with both arms raised above the head.[12] Many of these measures have been superseded by standardised units, but survive in an informal sense. We may have forgotten the *yepsen*, for example, a Middle English unit equivalent to a pair of cupped hands, but we still dole out food and ingredients by the pinch and mouthful.

Using our bodies to measure the world makes intuitive sense. It is a scale appropriate to human activity and means that measuring tools are always at hand. The same logic applied to many other pre-modern measures, which were defined by the expediencies of everyday life. This often meant they were elastic in their values, shrinking or expanding with their environment. Consider the old Finnish unit of length known as the *peninkulma*, which was originally measured as the distance at which a dog's bark could be heard[13] (around 6 kilometres). Such a unit would be imprecise, changing in length based on the type of land being measured (how far does a dog's bark travel through dense forest versus open valleys), but this flexibility offers its own store of information, giving some indication of terrain and accessibility. This aspect of measurement is most clearly demonstrated in medieval land units, which varied based on similar agricultural factors. The old Irish unit of the *collop*, for example, was defined as the amount of land needed to graze a single cow. Here is a unit that adapts to the practicalities of life, meaning lush, energy-dense pasture would be measured in smaller *collops* than the same area of barren hillside.

In the 1942 Irish novel *The Tailor and Ansty* by Eric Cross, the character of the Tailor uses the *collop* to demonstrate the wisdom of his forebears. 'An acre might be an acre of rock but you know where you are with a collop,' he says, disparaging a neighbour who boasts of owning 4,000 acres but has only 'enough real land to graze four cows'. To illustrate this ancient practicality, he describes a unit of time based on the lifespan of the rail, a family of birds that includes the moorhen and coot. 'A hound outlives three rails / a horse outlives three hounds / A jock outlives three horses / A deer outlives three jocks / An eagle outlives three deer / A yew-tree outlives three eagles / An old ridge in the ground outlives three yew-trees,' says the Tailor. No larger units of time are needed, he adds, as three ridges is 'as long as from the beginning to the end of the world'.[14] As the science writer Robert P. Crease has calculated, if a rail lives for a decade, then this reckoning puts the age of the universe at 65,610 years.[15] That's nothing compared to modern estimates of around 14 billion years but close to the estimates of the Middle Ages. As the Tailor says, the reason old units are superior is obvious: they were 'reckoned on the things a man could see about him, so that, wherever he was, he had an almanac'.[16]

The descriptive capacity of units had practical benefits, flowing to fit the contours of labour and land. Their use seems part of a world view that prioritised locality and tradition. But as society grew more interconnected, these measures created problems. If neighbouring regions used different units (or worse, the same units with different values), it stymied trade. Metrological dialects also reduced citizens' legibility, making it difficult for centralised governments to assess and tax the wealth of the people. Corruption, too, flourished with variable measures. Manorial lords, for example, would collect their feudal dues using capacity measures of grain larger than those used in markets and mills. When their dependents were cheated, what authority could they turn to? An absence of standardised measures created a power vacuum that was easy to exploit.

It was these factors that contributed to the single most significant event in the history of measurement: the creation of the metric system. This project, which took place during the final years of the eighteenth century alongside the French Revolution, was both symbol and instantiation of the era's politics. The metric system was designed by the country's intellectual elite, the *savants*, to reflect the era's ideals. They thought that the standardisation of weights and measures would eliminate some of the imbalances of feudal life, complementing the political equality of republicanism. To remove measurements from the vagaries of work and environment, they decided to base their units on what they considered to be an impersonal and incorruptible arbiter: the Earth itself. Many of the age's greatest intellects contributed to what would be a seven-year project to measure the planet and define the new unit of length, the metre. This would eventually be standardised as one ten-millionth of the distance from the North Pole to the equator, with the accompanying mass standard, the kilogram, equal to the weight of a cubic decimetre of water (that's a cube with sides one-tenth of a metre in length). These new units, as much the product of revolutionary tribunals as scientific calculation, would be universally accepted, said the *savants*, because they were abstracted from human affairs: impartial and unchanging.

These abstractions would, eventually, lead to the global adoption and domination of the metric system. As the historian Eric Hobsbawm has noted, metric units are in some ways 'the most lasting and universal consequence of the French revolution.'[17] By helping transform measurement from something particular to a specific time and place to something indiscriminately applicable, it allowed for organisation, analysis, and control on a scale undreamt of by our ancestors. And indeed, in the centuries since the metric system was created, it has been abstracted even further. The metre and its fellow metric units are no longer based on anything so crass as the span of the planet. No, they are now defined by constants of the universe itself, on the speed of light and spin of atoms, unchanging – so far as we know – throughout reality.

———

My journey into the history of measurement has taken me to many places, from the baking heat of Cairo, where the importance of metrology is catalogued in the stone of the pyramids, to the chilly museum in Uppsala that houses the first ever Celsius thermometer. But it was at the British Library in London that I did most of my work, arriving to read and learn from the countless academics and researchers whose work has enabled my own. Every time I walked to the entrance of the library, I would find my gaze captured by the building's guardian: a cyclopean statue of a man sitting on a stool, bending over to measure something on the ground in front of him. The figure's gaze is intent, his body huge and muscular, and in his outstretched hand he holds a compass that jags like lightning into the earth.

I thought the statue a good omen at first. Here was a monument to measurement, to its capability and authority, greeting me every day. But one afternoon, sitting in the courtyard during a lunch break, I idly looked up the monument's history and found a very different explanation of its meaning. The sculpture is from 1995, the work of Italian artist Eduardo Paolozzi, but is modelled on a 1795 watercolour by William Blake, which depicts Isaac Newton in an identical pose. Paolozzi intended the statue to represent a fusion of the humanities and sciences – to embody humanity's search for truth across different disciplines – but Blake had a far sharper critique to make: he was not celebrating the work of the great scientist but satirising his blindness. The figure in his watercolour is engrossed in his calculations, yes, but ignorant to the wonders of the world around him. He is hunched, not heroic; obsessed with his measurements and so missing entirely what his compass cannot contain.

This illustration happened to be one of Blake's favourites, which he reproduced many times over in his life, even tinting the colours of a final copy on his deathbed.[18] It represented one of the great intellectual struggles of his life: his reckoning with Enlightenment ideals of rationality and progress, which seemed to inspire and

William Blake's print of Isaac Newton depicts the great scientist as a divine geometer, unwilling to lift his gaze from his calculations

antagonise him in equal measure. Newton's pose in the watercolour recalls that of another compass-wielding figure in Blake's mythos: the god Urizen, whom Blake represented in his poetry and art as the first deity in existence; the embodiment of law, rationality, and order. (Although the name sounds biblical, it seems also to be a surprisingly candid pun on the words 'your reason'.) For the mystic and radical Blake, a man who spoke to angels and championed revolution, Urizen was not a noble presence but a repressive tyrant, who first weighed and measured the universe and in so doing imposed limits on the human soul. As Anthony Blunt, art historian and Soviet spy, put it: 'The effect of Urizen's creation [for Blake] was to crush man's sense of the infinite, and to shut him up within the narrow wall of his five senses.'[19]

This idea of measurement as repressive force was not what originally caught my attention when I was first enchanted by the redefinition of the kilogram. But as I waded deeper into the history of metrology, this aspect became impossible to ignore. Measurement is unquestionably a tool of control and, as a result, has been used throughout history to manipulate, persecute, and oppress. To measure something, after all, is to impose limits on the world: to say this far but no further. It means fitting reality into categories that can never capture its full complexity. For when we try to measure some particular aspect of the world, we are inevitably making a choice that reflects our biases and desires. Measurement is a tool that reinforces what we find important in life, what we think is worth paying attention to. The question, then, of who gets to make those choices is of the utmost importance.

These dynamics can play out in very different ways. Some examples of mismeasure may be petty and slight, like the humiliations of workplace bureaucracies – painful in the moment but forgotten easily enough. Others are difficult to fathom in the extent of their cruelty. Consider the horrors of eugenics or scientific racism: movements motivated by ugly notions of racial hierarchy but justified through the pretend objectivity of measurement, conveyed through comparisons of skull sizes and IQ tests.

For Blake and his sympathisers, such barbarism is an expected product of the civilised world. The twentieth-century philosophers Max Horkheimer and Theodor W. Adorno noted that the entire process of abstraction – of creating generalised rules and categories – is one of the foundations of modernity, for better or worse. It is a product of the same Enlightenment thinking that made Blake shudder and that continues to shape the societies we live in today. 'Classification is a condition of knowledge, not knowledge itself, and knowledge in turn dissolves classification,' they wrote.[20] The world as we know it today is 'ruled by equivalence'; by a desire to reduce everything to number and make 'dissimilar things comparable by reducing them to abstract quantities'.[21] Or, as the satirist Jonathan Swift put it centuries before:

> *Philosophers, who find*
> *Some fav'rite System to their Mind*
> *In ev'ry Point to make it fit,*
> *Will force all Nature to submit.*[22]

Perhaps this all sounds like rather a heavy burden to attach to measurement – to accuse tape measures of brutality and scales of injustice. But we regularly turn to measurement to fix the greatest problems in society, whether in healthcare, education, or policing, so why should we be surprised if it is capable of threatening our happiness too? This, I think, is the true beauty of the subject: its depth is hidden on the surface. Peel back that thin layer of familiarity and measurement is anything but banal. It is a complex and turbulent force that has shaped history; it has been a tutor to humanity and an overlord also. Over time, it has been the concern of gods and kings, and an inspiration to both philosophers and scientists. It is a child's art, practised with a pencil and ruler, but also the means by which some of humanity's greatest achievements have been orchestrated. In the final reckoning, measurement has left its mark on us all.

1

THE KINDLING OF CIVILISATION

---

*The ancient world, the first units of measurement,
and their cognitive rewards*

I have not diminished the palm measure
I have not falsified the cubit of land.
I have not added to the weights of the balance.
I have not nullified the plummet of the scales.

— OATH FROM THE EGYPTIAN BOOK OF THE DEAD[1]

Measuring the abundance of the Nile

We tend to think of measurement as something taken from the world: as knowledge extracted from nature by means of scales, gauges, and rulers. But this framing is just convention, and the opposite is equally true. The measure often precedes the measurement: it is the product of some complex system, perhaps unseen or not yet fully known, that exists prior to our attention and requires effort to be understood. In the same way that high-energy cosmic particles hurtled unscathed through the heads of our ancestors for millions of years before we designed the means to track them, we can think of measures as crowding the universe, innumerable and imperceptible, passing through us like spirits from another dimension. All we need to do is devise the means to know them.

In the world of ancient Egypt, one particular measure could be found long before people settled the land and would come to define the civilisation they built there: the bounty of the Nile, a liquid treasure metered out each year in floodwater and fertility.

'As Herodotus tells us, Egypt is the gift of the Nile,' says Salima Ikram, Professor of Egyptology at the American University in Cairo. She and I are bouncing along the streets of Cairo in a battered

Soviet-era Lada taxi, and Salima delivers this quote with a wink and a smile, aware of both the smoothness of her patter and the lazy demands of itinerant writers.

True to the clichés, as a first-time visitor in Cairo I feel utterly outmatched by the noise and heat of the city, but even more so by Salima herself, who is small, suave, and irrepressibly sharp. During our time together, she seems to field constant phone calls from friends and colleagues, organising dinners and archaeological digs with equal aplomb and addressing everyone she talks to as 'sweetie' and 'habibi'. When I ask if she's really on such friendly terms with all these people, she laughs and replies: 'Oh, no, it's just because I can't remember anyone's name.'

Herodotus was right about the Nile, of course: the river flows down from the south into Egypt from the highlands of present-day Ethiopia, flooding the surrounding plains with unusual regularity. For millennia, it clotted thick, rich mud on to the landscape every summer, and into this sticky terrain crops could be planted to ripen with minimal watering under the winter sun. By spring they would be ready for harvest, and as the mud dried and cracked open in the summer heat, excess water and minerals would drain away from the land like a bathtub with the plug pulled out, leaving the fields parched and ready for the cycle to begin again.

The ancient Egyptians were acutely aware of the Nile's importance and embedded its presence deep in their culture. The river's annual floods created the three seasons of their calendar: *Akhet*, or inundation, *Peret*, or growing, and *Shemu*, or drought. The floodwaters themselves were deified in the form of Hapi: an androgynous god depicted with a full belly and swollen breasts to signify the abundance he brought to the world.[2] It was the inscrutable will of Hapi that offered the best explanation for the Nile's largesse, and it was thought he released its waters each year from hidden caverns in the

24

mountains. As the floods cascaded down into Egypt, Hapi's spirit would flow across the land, trailed by a frolicking retinue of frogs and crocodiles. The enormous wealth created each year by the Arrival of Hapi nurtured a civilisation that has lasted for millennia. Even today, the Nile is indispensable to Egypt, providing 95 per cent of the country's water needs.[3]

Capturing this abundance required ingenuity, and Salima is taking me to see one of the tools used for this task: an artefact of ancient measurement that is testimony to metrology's role as the kindling of civilisation. It is a nilometer: a measuring tool used by the ancient Egyptians to gauge the depths of the Nile's floodwaters each year. These readings were vital, as the depth of flooding determined whether the year's harvest would be slim or bountiful, an insight that powered the operations of the state like the mainspring of a clock. If the nilometers said a famine was coming, then food would need to be set aside to sustain the people and stem unrest. If they predicted the harvest would be plentiful, then appropriate taxes could be levied in the form of crops, labour, and land. These resources then supported the business of the nation, contributing towards trade for silver or copper, rations for the army, and construction projects like the society-warping pyramids. Most importantly, the bounty of a good harvest would be used to fill the granaries that would feed the people in leaner years. But for all those decisions to be made, measurements had to be taken.[4]

Appropriately for this singular purpose, nilometers are simple things. In essence, they are just giant rulers, carved into columns, walls, or stairs that were themselves within reach of the Nile's waters. Each scale was marked in cubits, an ancient unit of length that was likely invented by the Egyptians before spreading to neighbouring cultures. The cubit is equal to the distance from elbow to fingertip (the English name is derived from the Latin *cubitum*, meaning 'elbow') and has maintained its familiarity today through widespread use in the Bible. Many different cubits have existed over the millennia, and the Egyptians themselves recognised two varieties: the 'common cubit' of six palms, and a longer 'royal cubit' of seven palms. Each palm is divided into four fingers, giving a total length for the longer cubit of roughly 52 centimetres or 20 inches.

As our taxi dodges in and out of traffic, Salima and I briefly burst from the streets on to a wide bridge that sails over the Nile itself, the air cooling as it streams over water. Salima explains that it's difficult to say for certain how the first nilometers were used, but they were undeniably 'a key feature of Egyptian life'. There were probably hundreds spread across the country, she says, and although we don't know when they were first introduced, it's her guess that they appeared around 3000 BC, when the first of Egypt's dynastic kings began to unify the lands.

'Even then people were reliant on the Nile, and one of the reasons we think the whole ancient Egyptian state came about, with the creation of writing and bureaucracy and so on, was to organise access to the water and land,' she says. 'You have to figure out a way of documenting who owns the water and who gets access to it, and that requires the state.'

Salima and I disembark on to Roda Island, one of several large islands in this section of the Nile. We ease ourselves out into the baking heat and approach the gates of the compound where the nilometer resides. This particular example is far from the oldest in Egypt, but it is one of the grandest. It's barely a millennium old and was built in the ninth century AD, when Egypt was part of the Islamic Abbasid Caliphate.[5] As Salima chats to a quartet of guards (who have sensibly abandoned their sun-drenched sentry post for a bench in the shade), I look over to the building housing the nilometer a few hundred metres away: a single-storey construction topped by a huge twelve-sided roof, the structure positioned at the leading edge of the island, pointing forwards into the Nile like the helmsman of a great ship.

Salima buys our tickets, and one of the guards ambles out towards the building ahead of us. It's mid-morning and we're the first visitors of the day. We stand about while he unlocks the door and then step gratefully out of the heat and down into the stony coolness below.

I'm shocked by what we find inside. For although the nilometer itself is plain – a single octagonal column built into the centre of a well roughly 11 metres deep – the roof above is dazzlingly ornate. Twelve symmetrical windows let light and air into the building, and the steeple-like ceiling is covered in a dense pattern of vines, leaves, and flowers, woven together like a verdant tapestry. The colour scheme is gold and green, and it feels like we're standing under the canopy of some mythical jewelled garden. Quietly, Salima slips the guard some money ('I'm keeping a tab, don't you worry,' she whispers to me), and he unlocks a private gate for us to descend into the well. Stone steps line the interior, but the absence of a handrail is unnerving, and I press my back flat against the wall for safety. Salima doesn't seem to notice, and meanders down in an unhurried fashion, stopping only to point out some of the more interesting graffiti to me. 'Look at this one,' she says, indicating carved letters on the wall. '1820, English tourists.'

Once we've reached the bottom of the well, I regain my equilibrium and take in our surroundings from this new perspective. It's dusty and damp, like the catacombs of an old church, and feels as much like a burrow as a basement. Looking up, I can finally appreciate the full glory of the twelve-sided roof, which fills the opening at the top of the well. Positioned just above the apex of the nilometer's column is a swirling circular motif that looks like a huge flower or a rising sun. I'm astounded by the beauty of the place. 'It's gorgeous,' I

The Roda nilometer was one of many used in ancient Egypt to measure the depth of the Nile's flooding each year and prepare the land for famine or feast

murmur, and Salima nods emphatically in agreement: 'It's absolutely fabulous.'

As the guard patrols behind the balcony above us, Salima and I chat in the shadows, standing either side of the tall marble measuring column, its millennium-old cubit markings visible as shallow grooves in the stone's surface. We don't know exactly how measurements from these tools were collated in pharaonic times, explains Salima; how frequently they were read or how long records were maintained. But it's clear that each nilometer had its recognised readings, both good and bad. In the first century AD, when Egypt was under Roman rule, historian Pliny the Elder described how figures taken from a nilometer thought to be in Memphis, the capital of ancient Egypt, were used to gauge the country's fortunes. 'When the water rises to only twelve cubits, [Egypt] experiences the horrors of famine,' he wrote. 'When it attains thirteen, hunger is still the result; a rise of fourteen cubits is productive of gladness; a rise of fifteen sets all anxieties at rest; while an increase of sixteen is productive of unbounded transports of joy.'[6] This figure of sixteen cubits seems to have significance beyond Pliny's recording, too. An eighteenth-century statue copied from a second or third century AD original that represents the Nile as a recumbent and muscular figure, for example, is surrounded by sixteen *putti*, or cherubs, each a cubit in height to symbolise the ideal flood.[7]

Such dedication to these figures may seem unusual, but the nilometers' readings were not mere trivia. Not only did they reveal whether the year would be spent in grim poverty or easy abundance, but they also reflected something of the country's divine and political fortunes. Religion and government were inextricably linked in ancient Egypt, with the country's priesthood serving as its civil service: overseeing laws, administering resources, and advising the pharaoh, who was himself a semi-divine figure. Spiritual ritual and bureaucratic rigour were therefore both needed to maintain prosperity, and the nilometers themselves were usually built inside temple complexes. The same priests who read these scales would oversee the religious festivals that celebrated the floods. 'The flooding of the Nile and fertility of the land were linked directly to the pharaoh's rule,' says Salima. 'So, if you get lots of bad floods, that means the gods are pissed off with the pharaoh, and, by extension, with all of Egypt.' In these circumstances, measuring the depth of the Nile seems more than a practical chore: it's a rite that gauges the favour of the gods.

As Salima and I stand for a moment in silence at the foot of the well, I nod to a shadowy tunnel in one side of the wall. 'Where does that go?' I ask. 'That', Salima replies, 'goes straight to the Nile. It let the waters into the well when the nilometer was in use. But it's blocked off now, thankfully. Shall we take a look inside?' We take out our phones and turn on their flashlights, shining them into the tunnel like explorers. As we shuffle into the darkness, edging away from the island's centre and closer to the Nile itself, I think about the mass of water flowing unseen over and around us, just metres away behind rock and stone. It's a force that has sustained this land for millennia – and a measurement waiting to be understood.

The invention of writing, number, and measure

Measurement was a crucial organising principle in ancient Egypt, but metrology itself does not begin with nilometers. To understand its place in human culture, we have to trace its roots back further, to the invention of writing itself. For without writing, no measures can be recorded. The best evidence suggests that the written word was created independently thousands of years ago by a number of different cultures scattered around the world: in Mesopotamia, Mesoamerica, China, and Egypt. But it's in Mesopotamia – present-day Iraq – where the practice is thought to have been invented first.

A brief sketch of the origin of writing goes like this: in the beginning there was the Thing, and the Thing needed counting. What the Thing was doesn't matter much. A flock of sheep, perhaps, or sheaves of barley: profits of the new system of settled agriculture, which had allowed cities with tens of thousands of occupants to appear for the first time in history. The women and men who dwelt in these cities wanted to keep track of their new wealth and decided to use clay tokens for the job. These tiny objects, the size of game pieces, were shaped as cones, discs, triangles, and cylinders and can be found scattered throughout the archaeological record like errant dice. The earliest date back to 7500 BC, in what would become the Mesopotamian civilisation of Sumer, home of the Sumerians.[8] The tokens seem to have been useful, as they multiply in form and number over the centuries. As city life became more varied in Mesopotamia, with inhabitants trading not only raw materials like wool and metal, but also processed goods like oil, beer, and honey, more tokens were created to represent these resources. Their appearance became more complex, with scratches added to their surface, adding a graphic element to their meaning. Fast forward a few millennia, and, like a shopper burdened with too much pocket change, the Mesopotamians were fed up with their clutter of tokens. To better organise them, they began making clay containers known as *bullae* to enclose them into groups. These *bullae* started appearing around 3500 BC, as bumpy spheres the size of tennis balls, filled with clay tokens and sealed like a baby's rattle. One *bulla* could then be used to track multiple items.

This technology had its advantages and disadvantages. If you are, for example, a Sumerian priest recording tributes from farmers, you'd be happy that your clay spheres couldn't be tampered with, but annoyed that you couldn't check their contents without breaking them. So, one day, while making your latest *bulla*, before you put the tokens inside, you press them firmly on to the clay's wet exterior

we've recovered only a handful of literary texts. Instead, the overwhelming majority of unearthed writing tablets – some tens of thousands – are administrative in purpose. These were composed by a class of professional scribes, who were 'the cohesive force that helped preserve and enrich' ancient Mesopotamia, fulfilling duties including 'temple functionary, court secretary, royal counselor, civil bureaucrat, [and] commercial correspondent'.[13] The library they created is one of receipts, contracts, shopping lists, tax returns, deeds of sale, inventories, wage slips, and wills. Over time, narrative writing like royal announcements and records of wars were added to the mix, but even these retain something of the catalogue format, listing provinces conquered, offspring born, and temples consecrated and desecrated.

There's some debate over whether this invention of writing enabled the first states to emerge, giving their rulers the ability to oversee and allocate resources, or whether it was the demands of the early states that in turn led to the invention of writing. Either way, the scribal arts offered dramatic new ways to process knowledge, allowing for not only superior organisation, but also superior thinking. Some scholars argue that the splitting of noun and number on clay tablets didn't just allow kings to better track their taxes but was tantamount to a cognitive revolution: a leap forward that allowed humans to abstract and categorise the world around them like never before.

Lists may not seem like cognitive dynamite, but their proliferation appears to have helped develop new modes of thought in early societies, encouraging us to think analytically about the world. 'The list relies on discontinuity rather than continuity,' writes anthropologist Jack Goody. '[I]t encourages the ordering of the items, by number, by initial sound, by category, etc. And the existence of boundaries, external and internal, brings greater visibility to categories, at the same time as making them more abstract.'[14]

Think about how spoken language tends to place information in a definite context. When recalling your day, you might say: 'I went to the shops and bought eggs, flour, and milk to make pancakes.' The list, by comparison, abandons continuity for atomisation, removing individual items from a wider narrative (to buy: eggs, flour, milk). It fosters what psychologists call 'chunking' – the process of breaking down large quantities of data into manageable subdivisions and measuring out the world in discrete packages. Most of us are aware instinctively of the benefits of this approach. When we're wracked by vague terror about tasks yet to be tackled, we often resort to list-making, paring down the madness of the world into something that can be managed one job at a time.

This categorisation of knowledge in early Mesopotamian society is evidenced by what archaeologists call 'lexical lists': tablets that simply list different classes of objects like the index of an encyclopedia. The exact function of these lists, which cover everything from types of trees to body parts and names of gods, isn't entirely clear. They might have been used to teach vocabulary or as practice for scribes, but what they show is ancient humans grappling with the problem of classification.

As Goody argues, the process of constructing a thematic list 'leads to increments of knowledge, to the organisation of experience.'[15] It is a precursor to organised philosophical systems, and, eventually, to science. Centuries later, in the fourth century BC, Aristotle would turn the list format into the bedrock of his thinking by divvying up all of reality in his great work, the *Categories*. This grand taxonomy draws many arcane distinctions: between the Eternal Mobile Substances (the heavens) and the Destructible Mobile Substances (the sublunary bodies); between the Unensouled Destructible Mobile Substances (elements) and the Ensouled Destructible Mobile Substances (living beings); and so on. None of the examples of this form prior to the ancient Greeks are anywhere near as philosophically complex, but they are elaborate and beautiful just the same.

One particularly famous example of the form comes from ancient Egypt and is dated to around 1000 BC: a product of the state's bureaucratic culture known as the Onomasticon of Amenopě. In its simplest form, the onomasticon is simply a list of some 610 entries: items that collectively span the known world. An introduction to the text states that it is to be used 'for instruction of the ignorant and for learning all things that exist: what Ptah created, what Thoth copied down'. It begins with the natural world: the first entry is 'sky', followed by 'sun', 'moon', and 'star', before proceeding through 'darkness' and 'light', 'shade' and 'sunlight', and tackling various earthly categories like 'river-bank', 'island', 'sand', and 'mud'. After describing the Earth, it moves on to its occupants, beginning with the supernatural – 'god', 'goddess', and 'spirit' – before progressing to the most important humans, starting with the royal court ('king' to 'queen' to 'king's mother'), then through high-ranking civil and military roles ('general' and 'deputy of the fortress'), and then on to the wider world of work. This is the most granular section, with several hundred entries offering a detailed picture of Egyptian society. It starts with the professional artisans ('sculptor', 'hour-keeper', and 'astronomer') before moving on to the lower orders ('steersman', 'herdsman', 'gardener', and 'dancer'). After the people have been dealt with, there's a section on the towns of Egypt, followed by types of building and terrain. After reaching the ground, we move on to survey its bounty: crops, vegetables, and other foodstuffs for over a hundred entries. The list ends when even these items have been broken down into their constituent parts, with the final three entries of raw meat, cooked meat, and spiced meat.[16] As the list's author has promised, we've been shown 'all things upon which Re [the sun god] has shone'; taken on a journey from the cosmic pantheon to the butcher's table in 610 easy steps.

The Egyptologist Alan Gardiner, who collated the various manuscripts that make up the text, was unimpressed. 'Certainly there was never written a book more tedious and less inspired than the Onomasticon of Amenopë,' he commented in 1947.[17] But three decades later, Goody finds much more value in the list, noting how the onomasticon demonstrates 'the dialectical effect of writing upon classification' to an unparalleled degree.[18] The entire text is a lesson in the power of hierarchy, as it blends together the spiritual and terrestrial realms into one great spectrum. Binaries like 'light' and 'darkness' appear in pairs in the list, accentuating their similarities and differences, while transitions between categories are observed with sensitivity. When 'dew' is listed in the onomasticon, for example, its placement mirrors the phenomenon itself: it appears on the border between earth and sky, a delicate imprint from one world to the next like the moisture that gathers on grass with the rising sun. Can a list be poetic? Can taxonomies do more than set their subjects in stone, but also enliven our awareness of them? I certainly believe so.

Thousands of years later, in an essay published in 1942, the writer Jorge Luis Borges captured the absurdity and scope of list-making with his own fictional taxonomy, supposedly found in an ancient Chinese encyclopedia titled *Celestial Emporium of Benevolent Knowledge*. In it, an unknown scribe orders all the animals of the world into fourteen categories. These include 'those that belong to the emperor'; 'trained ones'; 'suckling pigs'; 'mermaids'; 'those included in this classification'; and, my personal favourite, 'those that tremble as if they were mad'.[19] The divisions are precise, elegant, and incongruous. As the French philosopher Michel Foucault noted, the celestial emporium shows that lists require subtle thought; the ability to segment, categorise, and compare. These characteristics are a little hidden in ancient texts like the Onomasticon of Amenopě, but Borges hauls them to their feet and sets them dancing. As Foucault says: 'there is nothing more tentative, nothing more empirical (superficially, at least) than the process of establishing an order among things; nothing that demands a sharper eye or a surer, better-articulated language.'[20]

Divine time

The earliest applications of measurement are not simply concerned with organisational prowess. They are exploratory and curious: a method of interacting with the world that records its rhythms and guides our responses. As demonstrated by the nilometer's ability to gauge the favour of the gods, such measures have significance beyond the practical, especially in the ancient world, where the workings of nature are more frequently ascribed to divine causes. Because of this, early metrology is often a way of interacting with the supernatural, a connection seen particularly clearly in the origins of the measurement of time.

The fundamental unit of time is, of course, a day: the 24-hour period it takes the Earth to complete one rotation on its axis. As the Book of Genesis says: 'There was evening and there was morning, one day.' The value of this unit is nothing more than an accident of our planet's rate of spin, set when the Earth first coalesced from a cloud of gas and dust (and slowing down minutely ever since). But it's also a measure that is written into our biology, hard-coded in our DNA as circadian rhythms. This pattern of physiological events coordinates each of our bodies to the spin of the planet that is our home. It suppresses bowel movements in the night, raises alertness at dawn, and secretes melatonin at dusk to prepare us for sleep. Such rhythms exist not only in the animal kingdom, but in plants, fungi, and even some types of bacteria, a form of life so ancient that our evolutionary paths diverged between 1.5 billion and 2 billion years ago.[21] Of all the units of measure we have today, the day is the only one that had meaning before there was a mind capable of comprehending it. It's a unit that the earliest humans would have recognised and will be a measure humanity carries with it even if we leave behind the planet that created it. Astronauts on board the International Space Station, for example, which orbits the planet every 90 minutes, are certainly not bound by Earth's 24-hour cycle. But for the sake of their bodies, they replicate it by adjusting the intensity and colour of the station's lights throughout each 'day'.[22]

Days merely provide a short-term order, though, and it was the changing seasons, a result of the Earth tilting on its axis as it rotates around the sun, that gave us our longer units of time. As the climate shifts, it orchestrates not only the appearance of flowers, fruits, and crops, but the migration of animals and meteorological events like floods and monsoons, phenomena that any agricultural society must anticipate and respond to in order to be successful. A time to sow and a time to reap would have been known to the earliest farmers, and reading the cues of the natural world would have been obvious enough without the need for a standardised calendar.

As societies developed, though, more sophisticated timekeeping systems emerged, many of them based on the world's oldest scientific discipline: astronomy. As the seasons change, so does the night sky, with constellations appearing and disappearing and planets wandering across the heavens. The synchronisation of celestial and terrestrial changes naturally suggests a causal relationship, and so observing the stars became a way of explaining and predicting events on Earth. As archaeologist Iain Morley notes, it's in the calendars of ancient civilisations that 'natural and supernatural meet in explanations of the world.'[23] Take the Pleiades, for example, a small group of stars visible from many places on the planet that have been bestowed with mythological and spiritual import since at least 3000 BC. Various cultures have interpreted their movement across the sky as seven girls being chased by a bear, as a group of wives banished by their husbands, or a hen herding her chicks.[24] Except in the most southern latitudes, the Pleiades first become visible in autumn at dawn (an event known as heliacal rising), appearing higher in the sky each day until they reverse course in midwinter and disappear before the start of spring. For the ancient Greeks, these stars were the seven daughters of the Titan Atlas running from the hunter Orion while their father was occupied holding up the heavens. But they were also a signal to labourers: when the stars disappeared, it meant that the sailing season had ended and they must return to their fields. As the eighth-century BC poet Hesiod writes in his didactic poem *Works and Days*: 'When the Pleiades, the Hyades, and mighty Orion set, / remember the time has come to plow again – / and may the earth nurse for you a full year's supply.'[25]

For many ancient civilisations, there was a clear link between time-keeping and the divine. The earliest known monarchical dynasty in China, the Shang dynasty, which ruled from around 1600 BC, based its calendar on a ten-day cycle in which each day was associated with different spirits and dead ancestors. Sacrifices would be made to these figures in the hope they would intervene on behalf of the living, leading to a calendar that not only structured the present, but gave its followers a means of controlling the future as well.[26] In pre-Columbian Mesoamerican societies, this connection was most memorably expressed by the Long Count calendar of the Maya. Perhaps inspired by the repeating patterns of the natural world, these cultures framed time itself as a grand cycle – a 'cosmic odometer'[27] – that ticked over once every 1,872,000 days (5,125 years), resetting the universe in the process. This calendar has fascinated Western observers, especially as the most recent cycle was estimated to have started on 11 August 3114 BC and ended on 21 December 2012.[28] The flurry of excitement, paranoia, and opportunistic doomsaying that preceded this date was not echoed in contemporary Mayan communities, however. The ancient Maya were never particularly clear on what a reset of the Great Count meant, but it seems to have been more concerned with renewal than destruction. True enough, those hoping for the world to end in 2012 found that the sun came up once more on the morning of December 22nd.

These calendars were developed to meet both spiritual and practical needs, but they grew out of the cycles of the natural world, which is not so regular as it might first appear. At some point in the ancient past – perhaps even when hunter-gatherers were keeping unknown tallies on animal bones – the first significant calendar unit was formulated based on the cycles of the moon: the month. The moon waxes and wanes with convenient regularity, and ancient peoples defined the lunar month by measuring the interval between identical phases of the moon, some selecting the full moon as their starting point, and others the crescent or new moon. These changes in the moon's appearance would be announced by star-gazing priests, a practice from which we get the word 'calendar', derived from the Latin verb *calare*, meaning 'to call out'.

But while the lunar month fulfils some of our criteria for a successful unit, it is not perfectly consistent. And for it to be useful as a unit of measurement, a consistent value is required. The problem is that the length of the lunar month varies by more than half a day, with an average length of 29 days, 12 hours, 44 minutes, and 2.8 seconds. This means that when you are not simply marking each new month as it appears but creating a calendar that stretches years into the future, you quickly run into problems. You could create a written calendar with alternate months lasting 29 and 30 days, for example, but it will slowly go out of sync with actual observations of the moon, resulting in one full day's difference roughly every three years, a large enough effect to be noticed in a single lifetime. If your records and observational skills are good enough, you could ignore the moon altogether and base your calendar on the sun, replacing the period between phases of the moon with the period between equinoxes. But this also generates problems, as the average length of a year is 365.2422 days, meaning even if you add a leap day every four years, over the centuries your calendar will still go out of sync. Whatever method you pick, you're going to have to periodically add extra days to pad out the difference.

Ancient civilisations dealt with these problems in different ways. Some added extra months to their calendars at the discretion of priests or astronomers, while others kept multiple calendars: a lunar calendar for religious rites and a calendar based on the solar year for civic purposes. The ancient Egyptians were among those to maintain a dual-calendar system, with their civic calendar comprising the three 120-day seasons based on the flooding of the Nile. This only gives a total of 360 days for a year, so an additional five 'epagomenal' days were added after the twelfth month to bring the total up to 365[29] – a practice copied in a number of cultures.

These extra days are curious phenomena, drifting unmoored from the regular schedule of normal life. In ancient Egypt, they were likely given as a holiday to labourers,[30] a concession offered not out of generosity but caution. As a transitional period between the old year and the new, the epagomenal period was considered to be perilous for the soul. Priests performed special rituals to safeguard the land; men and women made magic charms for their protection; and everyone, it seems, trod lightly for fear of disturbing some vast and unknown evil.[31] Not coincidentally, this tradition appears in the Mayan civil calendar too, which was separate to that of the Long Count and contained 18 months of 20 days, with five monthless, or *Wayeb*, days added to the end of each year. During this time, the 'portals between the mortal realm and the Underworld dissolved', letting evil spirits slip into the world to cause mischief.[32] Again, rituals and spells were carried out to safeguard the people. All a result of a calendar that couldn't quite capture the irregularities of the solar system.

For many in the West, we still experience a feeling of temporal suspension in that five-day period between Christmas Day and New Year. Are you at work or at rest? Are you prepared for the next year? Do you even know what it is you need to be prepared for? The calendar may be a human creation, an attempt to derive structure from the natural world, but like all measurement it creates its own realities too. Metrology helps to organise our lives, and as a result we imbue these systems with gravity and power, making it all the more important to understand the influence they have.

Close at hand: the first units

In the *Epic of Gilgamesh*, a 4,000-year-old Mesopotamian text that is a notable exception to the list-writing culture of the age, one particular scene illuminates the practice and limitations of early measurement.[33] It comes towards the end of the story, which is itself the first known tale of a tragic hero, as the eponymous Gilgamesh – a superhuman Sumerian king who is two-thirds god and one-third man – tries to avoid the fate that awaits all living things and attain immortality.

To discover how to cheat death, Gilgamesh travels to the home of Utnapishtim, a mysterious being known as The Faraway. Utnapishtim was once a mortal man, but he and his wife were granted eternal life by the gods after saving a selection of humanity from a world-destroying flood. (This is just one of many similarities the text shares with later biblical tales.) When Gilgamesh reaches Utnapishtim's home, he asks how he, too, might avoid the curse of death. The Faraway responds by offering a simple test: if our hero can stay awake for six days and seven nights, then Utnapishtim will teach him the secret of everlasting life. Gilgamesh readily agrees, but as soon as he sits down to prepare for this trial, his long adventures begin to catch up with him. 'A mist of sleep like soft wool teased from the fleece' drifts over his mind and he falls into a deep slumber. Utnapishtim is unsurprised and tells his wife to record how long Gilgamesh sleeps by marking the wall and baking a loaf of bread each morning to place by the demi-god's head. As the days pass, the bread is baked, and Gilgamesh sleeps on. Until 'there came a day when the first loaf was hard, the second loaf was like leather, the third was soggy, the crust of the fourth had mould, the fifth was mildewed, the sixth was fresh, and the seventh was still on the embers. Then Utnapishtim touched him and he woke.'

On waking, Gilgamesh insists that he was only resting his eyes, that of course he didn't fall asleep, how could he, such a big strong man? But when he sees the line of loaves by his bedside, a spectrum of decay, he can deny it no longer. The admission is catastrophic. If he cannot defeat sleep, what hope does he have of ever beating death? 'What shall I do, O Utnapishtim, where shall I go?' he laments. 'Already the thief in the night has hold of my limbs, death inhabits my room; wherever my foot rests, there I find death.'[34]

The mouldy bread brilliantly illustrates the inescapable nature of Gilgamesh's mortality, but it also offers a tantalising glimpse of ancient measurement. Note that although Utnapishtim's wife marks the days on the wall, it is the loaves that provide incontrovertible proof of the passage of time. There are no clocks or calendars to corroborate events, but the testimony of the natural world cannot be denied. The bread also demonstrates that when people need to measure some aspects of the world, to quantify it and record it, they often turn to the actions and objects of everyday life. It's a method that is paradigmatic of the improvised measures of early metrology.

The science writer Robert P. Crease suggests that there are three crucial properties that units of measurement must possess: accessibility, proportionality, and consistency.[35] Accessibility is needed because you can't measure something if you can't find your measuring standard. Proportionality is necessary because no one wants to measure mountains with matchsticks. And consistency is perhaps the most important attribute, as a unit of measurement that varies unexpectedly simply can't do its job (though one that flexes with intention can be useful). The loaves baked by Utnapishtim's wife more or less fulfil these criteria, although the 'unit' is technically the rate of decay, rather than the bread itself.

Creating a reliable bread clock of our own would make for a tricky week in the kitchen. We'd need to follow a standardised recipe when making the dough, and ensure that the baking conditions and room temperature were identical to minimise variations in the decaying process. Luckily for early societies, there were better sources of measurement than the hearth. When it comes to measuring the physical world around them, the single greatest resource of units of measurement for humans has been one that's very close at hand: our own bodies. Every culture that makes measures has deployed the body in this way, and the resulting units neatly fit Crease's criteria. They are accessible, always at hand; appropriate for measurement at an intrinsically human scale; and consistent (more or less) from one person to another. It's no surprise that just about every culture has some sort of unit of length based on the foot (and normally called just that), while usually taking units from various parts of the upper body and arms. Many of these ancient units are still familiar today, including the cubit, fathom, span, and hand.

While the body accounts for the bulk of small-scale units in ancient cultures, there are other regular sources. If you're casting around for accessible, proportionate, and consistent units, then seeds, for example, are perfect. Poppy, millet, and wheat seeds have all been used to create measures of length and weight, some of which are still in use today. The barleycorn, for example, has a long history in Great Britain as a unit of length. It's equal to a third of an inch, or around 0.8 centimetres, and has been associated with length at least since the early fourteenth century, when King Edward II declared that 'three grains of barley, dry and round make an inch'. This definition later became standardised in the imperial system of measurement, and is still in use today as an increment in UK and US shoe sizes. The difference between sizes is equivalent to a third of an inch, which shoemakers call a barleycorn.[36] The other familiar botanical unit of measurement comes when seeds are used to measure mass, not length. The grain has long been the smallest unit of weight for various English measures based on the pound (it's since been standardised as 64.79891 milligrams), and we still measure precious gems like diamonds and emeralds with the carat,[37] a unit derived from the seed of the Middle Eastern carob tree (and that is now standardised as 200 milligrams).

Perhaps the most imaginative form of everyday measures, though, are those used to estimate distances beyond a body's span. Before humans had access to maps or basic surveying tools, describing the distance between far-flung locations required ingenuity. A small number of cultures measured longer distances in multiples of body length, but such units are disproportionate (we don't often measure the distance between cities in metres or feet). So, they devised longer units that required less multiplication, sacrificing accuracy as a result. A bowshot, a stone's throw, and an axe-throw are all common measures of length in early civilisations, as are units based on sound, such as the distance from which you can hear someone shout or a dog bark. Longer units are even more imaginative, and are often drawn from acts of consumption. The indigenous Ojibwe people of Canada, who travelled by canoe in summer and snowshoe in winter, measured distance by the number of pipes of tobacco one might smoke on any given journey, and would comment of a particularly short voyage: 'My pipe had hardly burned out when I reached the place.'[38] Inhabitants of the Nicobar Islands in the Indian Ocean, meanwhile, gauged voyages by the number of coconuts drunk during their duration. To travel from one side of an atoll to another might be a length of seven or eight coconuts, a unit that offers information not only about distance, but also about the supplies needed for the journey itself. As with the measurement of time, people have often looked to the consistencies of nature to create such predictable units. The Saami cultures of northern Europe have a unit of distance known as the *poronkusema* that's equal to around 6 miles. It means, roughly speaking, 'reindeer's piss', and measures the distance a reindeer can walk before urinating. It's a delightful example of emergent measurement: a unit that springs from a specific animal which defines a people's way of life. For centuries, Saami have relied on reindeer for food, trade, and even labour. It's only natural that close company led to close attention, one day producing the observation – perhaps as a joke at first, then a habit – that, boy, those reindeer sure do piss on the regular.

Units like these seem exotic or fantastical to us now, artefacts of a time when the world thrummed with variety and there were truly authentic ways to be. But these sorts of ad hoc estimations have never gone away. Still today we gauge distance through contextual measures of time and activity. We explain to a friend that the next pub is just a five-minute bike ride away, or that the beach is just an hour's drive. We improvise new units, too, estimating our walk to work in podcasts, or telling ourselves that a flight is only three movies long. Such measurements are useful because they transfer information from an objective realm of distance to a subjective one of experience. They allow us to contextualise the world around us and make sense of it, just like the loaves of bread used to convince Gilgamesh of his slumber.

———

To harness the full potential of units of measurement, we need some degree of consistency. Subjective measures contain useful knowledge, but this can be difficult to share, based as it is on an individual's experience. The historian of science Theodore M. Porter describes measurement as a 'technology of distance':[39] a tool that uses shared rules to bridge disparities of culture and geography, allowing for the exchange of information. Following this logic, if measurement is another sort of language, then just as with words, individual units need reliable definitions in order to communicate.

In the earliest societies, these definitions might need to be shared no more widely than a single settlement, meaning that the definition can be no more complex than knowing that a unit was equal to a certain body part. As these communities grew in size, though, the inadequacies of this system would become evident. Imagine a disagreement over trade or tribute: it is a moment of friction that demands greater consistency in measurement.

Evidence of this tension can be seen in societies attempting to wring greater precision from the vagueness of anthropomorphic measures. Ancient Chinese texts distinguish between units of male and female body parts,[40] for example, while it was Ethiopian folk wisdom to ask a friend with 'long arms' to go to the market on your behalf, the better to extract favourable measures.[41] Sometimes these accommodations would be enshrined in law, as with the proclamation of King David I of Scotland in around 1150, in which he defines the inch as a thumb-length 'mesouret at the rut of the nayll', with the addendum that this unit needs to be taken as an average of 'the thowmys of iii men, that is to say a mekill [large] man, and a man of messurabel statur, and of a lytell man'.[42] By combining measures from three men – little, medium, and mekill – the law accounts for vagaries in anthropometric units, even if the result is still essentially a rule of thumb.

In Cairo's sprawling Egyptian Museum, you can find the means by which measures were first made consistent in ancient Egypt. They are cubit measuring rods, among the world's first standardised units of measurement. They were constructed from stone and wood, with intervals marked out in palms and fingers and their lengths certified by a central authority for official use.

The importance of these measures for ancient Egyptians is clear not just from tools like the nilometer, but from the sophistication of the country's agricultural and architectural achievements. Each year, during the inundation of the Nile, the floodwaters would destroy the boundaries of the river's surrounding farmland, and it was the job of a specialist corps of surveyors, known as *harpedonaptae*, or 'rope-stretchers', to restore order to the land. Using knotted ropes pulled tight to avoid sagging, they would venture into the mud and redraw the boundaries of the fields, ensuring that the waters unleashed by the river could be put to productive use. Their work was one of coordination and communication: minimising disputes between farmers and ensuring that the productive land was not wasted. The same ropes were also used to mark the foundations of buildings created from this wealth: to lay out plans for temples, tombs, and, of course, the great funereal pyramids that have come to dominate our understanding of the ancient Egyptians.

Visiting the pyramids of the Giza Necropolis in Cairo today is a lesson in the deficiencies of the imagination. We know the images of these monuments only too well, repeated throughout books and films and posters, but they are still boggling to behold in person: so massive that they seem like permanent fixtures of the landscape, so obviously artificial that they constitute a direct address to the spectator, a message from our ancestors, millennia old: 'We were here, and you are here, but *we are still here too*.'

The coordination of labour and resources needed to create the Great Pyramid is as awe-inspiring as the thing itself. Tens of thousands of workers toiled over decades to create the bigger pyramids, setting up towns just to support themselves, with bakeries and kitchens to feed the craftsmen, dormitories for sleeping, and graveyards to bury them in.[43] The American historian and sociologist Lewis Mumford has argued that it's this sort of mass organisation that actually sets humanity apart from other species, not individual displays of intelligence or mastery of any particular tool. 'Neither the wheeled wagon, the plow, the potter's wheel, nor the military chariot could of themselves have accomplished the mighty transformations that took place in the great valleys of Egypt, Mesopotamia, and India,' writes Mumford.[44] Instead, it was concepts of order, like 'the abstract mechanical system' and 'exactitude in measurement',[45] that enabled this 'vast explosion of power'.[46] Mumford refers to the resulting societal order as the 'megamachine' – a system that resembles machines as we now know them, but that is composed of human, not mechanical, parts.

The cubit measuring rods found by archaeologists reinforce this link between measurement and civilisation-building. Take, for example, the tomb of Kha, a respected architect who served three successive kings during the Eighteenth Dynasty (around 1400 BC) and who earned, for his trouble, a rich burial alongside his wife Merit. The couple's tomb, a rare find discovered in 1906, seems to have been prepared during Kha's lifetime and packed with hundreds of items to accompany the dead into the next life. These included furniture, food, and toiletries, but also a pair of cubit measuring rods, of very different design and significance.[47]

The first rod is a ceremonial item, exactly the sort of thing you'd expect to find in an ancient Egyptian tomb. It's finely crafted and covered in gold, with hieroglyphic engravings noting that it was presented to Kha by Pharaoh Amenhotep II, no doubt in recognition of services rendered. The second, by contrast, is plain but functional. It's made of a heavy, maroon-coloured hardwood that's been polished to a shine, with interval markings bearing the remains of a white pigment that would have shown up clearly against the dark-brown background. The rod has a hinge, folding in the middle for easy transportation, and was found tucked inside a soft leather satchel complete with carrying handles.[48] All the evidence suggests it was Kha's personal ruler during his lifetime, so important that he could not bear to be parted from it in the afterlife. Laid next to its ornate twin, the two measuring rods demonstrate the importance of measurement to the ancient Egyptians. Together they are a pair of magic wands, one wood and one gold, representing craft and authority – the means by which Kha raised temples and tombs from the desert, including his own.

2

MEASURE AND THE SOCIAL ORDER

———————

*The importance of metrology for early states
and the fabric of society*

A just weight and balance are the Lord's: all the weights
of the bag are his work.

—PROVERBS 16:11 (KJV)

Part standard, part sceptre

In the middle of the Louvre Museum in Paris, in the department of
Near East Antiquities, you can find an ancient reminder of the power
of measurement. It is the headless statue of a man named Gudea, a
Mesopotamian prince who ruled the powerful city state of Lagash
between the years 2144 and 2124 BC. The statue is carved from a speck-
led black rock known as diorite that's been polished to a deep and
starry sheen. The prince is seated and barefoot, dressed in a simple
cloak, with his hands clasped in front of him in a bearing of piety. His
fingers are thin, straight, and perfectly spaced like the teeth of a comb,
and in his lap rests his legacy: architectural plans for a new temple and
a marked ruler with which to build it.[1] The ruler itself, one of the earliest depictions of such a measuring instrument in history, is a long tri-
angular prism; a brutalist Toblerone with regular incisions cut into its
two visible faces. It is the symbol of Gudea's achievements. He wanted
to be remembered as a pious man and a builder of temples – 'a task
for gods and kings' in the ancient world, as one historian puts it.[2] The
temple was the central unifying institution in these early states, and
its plans were often thought to be divinely bestowed. Their execution
was, therefore, proof that the mantle of power had passed to the right
individual,[3] and so, like the Egyptian architect Kha, Gudea evidently
thought it fitting to take his measures with him into the afterlife.

Escape the crowds at the Louvre and head to the other side of the city, to the industrial design museum, the Musée des Arts et Métiers, and you can see how this connection between measurement and political authority survived well into the nineteenth century. There are no statues here, but standard units of weight, length, and capacity that jostle for space in busy display cabinets. Some are plain and functional, but most are elaborate affairs: ornately decorated jugs and lumps of metal stamped with crests and coats of arms. In one tall glass-fronted case you can see a collection of length standards from across the centuries: metrological swagger sticks fashioned from wood, bronze, and steel. Like Gudea's ruler, they are hybrid objects, part measuring standard and part sceptre, and proof of the utility of measurement in constructing social order. From the ancient world onwards, measurement has been embraced not only for its practical benefits – for its utility in tasks like construction and trade – but also for its ability to create a zone of shared expectations and rules; to mediate our experience with the world and with one another, ensuring that interactions between two strangers who live under the same set of measures can be validated and trusted.

It's often assumed that the state is needed to create this shared space, but historical evidence of the development of measurement suggests otherwise. Take, for example, the ancient use of mass standards: stones carved into regular shapes like spools, cubes, and ovals that were placed in balance pans to weigh goods. These can be found buried deep in the archaeological record, appearing from around 3000 BC onwards. This is centuries before the first descriptions of 'royal' standards appear, suggesting rulers co-opted as much as they created consistent standards of measure. Despite the lack of any central regulation, these ancient mass standards are incredibly consistent in their values. One analysis of more than

2,000 standards used across Mesopotamia, the Aegean, Anatolia, and Europe found that the weight of these stones differed very little between 3000 and 1000 BC. The total variation among the standards, which were recovered from locations thousands of kilometres distant, is between just 9 and 13 per cent. The conclusion is that Bronze Age merchants were capable of regulating units of measurement without the need for an overarching authority, with each individual meeting between traders serving as an opportunity to compare and adjust their weights.[4] Consistency here is a product of communal observance as well as the proficiency of the measuring tools being deployed. Unlike some other forms of measurement, weighing objects in this era could be achieved quickly and accurately through the use of an equal-arm balance – a tool so simple and complete in its design that it's still in use, unchanged, in many parts of the world thousands of years later.

However, just because measurement can be regulated without centralised authority doesn't mean rulers have ignored the potential of these systems. As the historian Emanuele Lugli has noted, units of measurement are, for the powerful, 'sly tools of subjugation'. Each time they're deployed, they turn the world 'into a place that continues to make sense as long as the power that legitimises the measurements rests in place'.[5] In other words: measurement does not only benefit from authority – it creates it too. As a result, the regulation of units of measure has been embraced by various political systems over the millennia. And from the ancient world through to the early modern nation state, enforcing these reliable units has been both a privilege and a duty, as necessary in justifying a leader's rule as the punishment of criminals or maintenance of roads.

Warnings against cheating with false measures can be found in some of the earliest known legal records, such as the Code of Hammurabi, authored around 1750 BC. Among its infamous prescriptions of an 'eye for an eye' justice, the code rules that wine-sellers who give customers short measures should be 'thrown into the water' (a euphemism for drowned).[6] Even the stele on which the laws were carved and first discovered is surmounted by a relief showing Hammurabi being handed a ceremonial measuring rod and rope by the Babylonian sun god Shamash – symbol of his authority. Millennia later, in thirteenth-century Europe, the regulation of measures was still taken extremely seriously. The Holy Roman Emperor Frederick II mandated that for a first offence of cheating with false measures the perpetrator gets an on-the-spot whipping, for a second one of their hands would be cut off, and for a third the punishment was hanging.[7] And failure to regulate units of measurement could destabilise regimes, as illustrated in England in 1215 with the signing of Magna Carta, through which the reluctant King John made a number of concessions to his unhappy barons. These included the promise that there would be but 'one measure of wine throughout our whole realm, and one measure of ale and one measure of corn [...] and one width of dyed and russet and hauberk cloths [...] And with weights, moreover, it shall be as with measures.'[8]

The importance accorded to measurement in these documents can be explained in part by the discipline's necessity for social cohesion. Most petty crimes affect only individuals, but false measures can spread distrust through an entire community. Talmudic law recognises this by noting that while many crimes can be repented for, no one can fully repent for cheating with measures as they can never account for the full impact of their actions. As soon as the crime is committed, it spreads like slander, eroding trust and fomenting suspicion. 'The punishment for unjust measures is more severe than the punishment for immorality, for the latter is a sin against God only, the former against one's fellow man,' says the Mishneh Torah.[9] Measurement is a covenant that binds communities together and is therefore impossible to fully account for at the level of the individual. This also suggests why regulation among community members, as with the example of the Bronze Age weights, can be as effective as a vertical approach: everyone has a stake in ensuring fair units are used.

Sometimes, the connections between measurement, political authority, and social order can surface in particularly surprising ways. In early imperial China, for example, the practice of metrology was closely tied to the music of imperial court rituals, which were themselves crucial in directing the flow of power among bureaucrats and the nobility. The link between measurement and music can be traced back to stories of the mythical Yellow Emperor, Huangdi, who was said to have created the first musical pitch pipes when he commanded his music master to cut bamboo stalks to specific lengths, matching the cries of male and female phoenixes.[10] These pitch pipes, known as *lülü*, defined the harmonic parameters of traditional Chinese music, and were used to tune the instruments of the imperial court. As a result, their exact value was not simply a matter of aesthetic importance but held 'cosmic significance', connecting the rule of the emperors to a semi-divine past.[11]

Because the pitches of the *lülü* were determined by their length, the value of the units used to measure the pipes could become a battleground for political factions. This dynamic is seen most clearly in the life of Xun Xu, a senior court official in the third century AD. Xun Xu was tasked with reorganising the imperial state led by the newly inaugurated Jin dynasty,[12] which controlled the south-eastern portion of present-day China. He sought to legitimise the rule of his new master, Emperor Wu, through the oblique politics of the imperial court, and by enacting a number of reforms to strengthen the emperor's authority. These included changing the basic unit of linear measure, the *chi*. To define the new unit, Xun Xu raided tombs from the ancient Zhou dynasty, a long-lasting lineage that introduced many of China's enduring political and cultural traditions. He dug up old Zhou jade rulers and used these as a template for his new Jin measure, which in turn altered the pitch of the *lülü*. He argued that by doing so, Emperor Wu was restoring the wisdom of this earlier, hallowed age, and that Wu's predecessors had been literally out of tune with the harmony of the ancients. The Jin reforms – both metrological and musicological – would restore this glorious past.

For Xun Xu, though, this meddling with measures did not have a happy end, and when he debuted his newly tuned instruments at court, the reception was not wholly harmonious. When the band started playing, the respected scholar and musician Ruan Xian, who was part of a group aligned with the pre-Jin regime, complained that the resulting harmonies were too high in pitch. 'A high pitch connotes grief,' Ruan is said to have commented. 'These are not the pitches of a flourishing state, but the pitches of a dying one. The music of a dying state is sad and full of longing, and its people are full of misery.'[13] His musical augury proved reliable. Not many years later, Emperor Wu died and the Jin state was plunged into disorder as eight rival princes struggled for power. Finding harmony among so many competing standards would be too much even for Xun Xu's cleverness.

Elastic measures

As the example of Xun Xu suggests, early metrology is defined by its plurality, with units developing in different regions and trades to meet the needs of the moment. There are periods of unification, certainly; eras of political stability that lead to a smaller number of measures embraced across a larger area (see, for example, the rise and fall of the Roman Empire, and the lasting influence its units had on the territory it governed). But a general rule is that prior to the ascendancy of the metric and imperial systems in the nineteenth century, the dominant theme of metrology is variation. However, the mistake is assuming that such diversity is without order.

In his pre-eminent study of metrology in medieval Europe, the Polish historian and sociologist Witold Kula shows that the abundance of early measurement comes with its own, complex rules that reflect the realities of life and labour. Measures in this period, he writes, 'are not as inaccurate as they may at times appear to us and the differences in them, as well as the coexistence of different methods of measuring, have a profound social significance.'[14] Without direct political oversight, small towns and villages could still enforce fair dealings in measurement through communal censure, says Kula, but the most notable quality of the units themselves was their elasticity. Many measures changed in value depending on how and where they were applied, reflecting the needs of the people and society that used them.

This elasticity was not the norm, by any means, but it was true for many important early units of measure. Consider, for example, the measurement of land in the Middle Ages, when the most common units were defined by the area that could be ploughed in a single day. By measuring the land in labour rather than physical area, these units adapted to the quality of the terrain. Harder soil or uneven ground would be measured in smaller units because it took longer to plough. As a result, the act of measuring encoded geographical and agricultural information relevant to workers, with such units found across Europe in this era. German-speakers used the *Tagwerk*, or day's work, equal to about 3,400 square metres or 36,600 square feet, and subdivided it into a *Morgen*, the area of land worked only in the morning. (Interestingly, the *Morgen* had a value of around two-thirds of the *Tagwerk*, suggesting that the early-rising farmers got most of the day's work done before lunch.) In Italy the equivalent was the *giornata*; in France the *journal*; and in Russia the *obzha*, which again reflects local agricultural practice as it was calculated from a day's ploughing done by horse, not ox. In many areas, these units of land-labour were even specialised to suit specific types of agricultural work. In the wine-growing Burgundy region, for example, the *journal* was only used to measure the size of fields growing grains, while vineyards were measured by the much smaller *ouvrée*. The difference in size between the *ouvrée* and the *journal* reflects the slower, more intense pace of work when one is tending vines compared to ploughing fields.

In England, we can see how units emerge from other practicalities of work. There's the furlong, for example, derived from the Old English for furrow, *furh*, and long, *lang*, which measures length, not area, and refers to the distance that could be covered by a team of oxen before they needed to rest (and which is now standardised as 201 metres or 660 feet). Furlongs were used to calculate the day-work unit of the *aker*, from the Old English *aecer*, meaning 'open field' (and from which we get the modern acre of 4,047 square metres or 43,560 square feet). Originally, an acre had a rectangular shape, one furlong long and one chain across (that's 20 metres or 66 feet). This was the result of farming practices that divided land into long, narrow strips, ensuring that fields could be lined up against a river-front so that each individual plot had access to water. The rectangular units also reflected the practical difficulties of turning a team of slow, heavy oxen around to plough in the other direction. Curiously, many of these land-work units seem to inflate in size as the centuries progress, which perhaps reflects an increase in agricultural efficiency. As farmers developed better methods and equipment, they were able to work greater areas of land in any given day, and so their day's work – and its unit of measure – grew in response.[15]

Other times, land units varied not in relation to working hours but to seed capacity. In Bourges in France in the eighteenth century, land was measured by the *seterée*, the area that could be sown using one *setier* of seed, a unit of capacity.[16] Fertile areas would be measured in smaller *seterées* and poorer ones with larger units, as seed could be sown more thickly on high-quality soil that could support more crops. In general, these variable units stayed in use at least into the nineteenth century, and their appeal is easy to understand. They're rich in information that modern measures just don't capture, shrinking and expanding to suit the particularities of their environment, like the quality of the soil or evenness of the terrain. Understanding how much land could be ploughed in a day or how much seed was needed to sow the next harvest was essential knowledge in an agricultural economy, and the rise and fall of these measures follows the contours of economic and industrial development. A study of the use of land-seed measures around Pisa in Italy shows that the unit survived for longer in rural communities where agricultural activity was dominant and disappeared more quickly in urbanised areas, to be replaced by inelastic units[17] (what we might think of as modern measures). As Kula notes, many historical studies of medieval metrology refer to the 'primitivism' and 'crudity' of elastic units, but in reality they are well fitted to the needs of the people who used them, embodying the relationship of humans to the land and capturing the necessities of their work.

———

Elastic units of measurement were not always useful to peasants, though, and were often exploited by the rich and powerful. A good example of this can be found with one of the most notorious measures of the Middle Ages: the grain measure. These units were usually gauged by capacity rather than weight, which is where we get familiar units like the bushel, containing around 8 gallons or between 35 and 36 litres of fluid (depending on whether you're measuring in imperial or US customary units). The word 'bushel' appears in the English language around the fourteenth century, and comes from the Old French *boisse*, another grain measure that was itself derived from the medieval Latin for a 'handful', or *bostia*. As with units of land, each country in Europe had a number of different capacity measures for grain, which were themselves often differentiated by crop (with more valuable grains like wheat measured with smaller units than cheaper alternatives such as oats). But, as anyone who uses American cups as a unit of capacity in the kitchen knows, the tricky thing with measuring dry goods by capacity is that the space they occupy can change depending on how you fill your container. Shake a cup of oats from side to side and the grains will settle, allowing more to be added. In other words: the method makes the measure.

This might not sound like cause for scandal, but imagine for a moment that you're a peasant in the Middle Ages. The grain you get from your farm is needed not only to feed your family, but also to pay your feudal dues and barter for goods at the market. When you take it to be sold, you are watching someone measure out not only months of labour, but, potentially, the future of your family. In lean years, shaking or striking the container to let the seeds settle could be the difference between survival and starvation. As a result, the activity of measuring grain is one of the most intricately controlled aspects of metrology in this period, with countless laws passed to regulate its measurement. These included ruling on whether grain should be poured from shoulder height or 'dropped-arm height' (the latter allowing the substance to compact more tightly); whether the measure was to be shaken after being filled or its contents pressed down, and whether the grain was to be measured 'heaped' or 'striked' (that is, whether it was allowed to pile up above the top of the container or levelled off with a special stick known as a strickle). These were not minor issues, either. A heap could account for a third of a bushel's total size when measuring wheat and rye, for example, and added as much as 50 per cent extra capacity to the units for grains like oats that pile more readily.[18]

This awareness of measurement was so ubiquitous that these acts became proverbial. One Polish motto warns against success by comparing it to methods of grain measurement: 'Stick your head out of the bushel and you'll get it levelled by the strickle.'[19] And in the 1611 King James Bible, Jesus promises attendees at the Sermon on the Plain that whatever good they do in life will be repaid with a 'good measure, pressed down, shaken together, and running over'.[20] In other words: unlike those tight-fisted lords, God is generous with the grain, offering heavenly rewards in abundance rather than making you fight for scraps. And warnings to watch the method of measurement closely last long into the modern era. A farming manual printed in 1846 advises readers to avoid selling grain to millers at the site of their mill as the 'continuous shaking of the building' caused by the machinery will 'ensure that the grain being poured into the measure will be most tightly compressed'.[21] Such practices were common beyond Europe too: one study of peasant economies in Burma in the 1920s includes an anecdote about a landlord who collected payment from his tenants using a basket so much larger than usual it was nicknamed 'the cart breaker'.[22]

Each moment of metrological imprecision offers the powerful and unscrupulous an opportunity for profit. Combined with the inequality experienced by serfs, whereby manorial lords could act with near impunity to extract as much labour and goods as they liked from tenants, it created widespread exploitation. The most common infraction seems to be nobles and merchants insisting on using their own set of capacity measures when taking payment, which presumably offered unfavourable portions compared to those used among the peasants. One of the most well-documented records of this sort of manipulation can be found in the *Cahiers de doléances,* an eighteenth-century survey of complaints and grievances collected in the run-up to the French Revolution. Issues of measurement figure heavily in the complaints of the peasantry, who demand that their lords be stripped of authority over weights and measures. One parish complains that 'the unhappy tenants are overburdened not only by the excessively large measures but also by the ruthlessness with which the payment in kind for poultry is levied', while another demands that 'all landowners, be they nobles or bourgeoisie or common people, be henceforth obliged to accept [...] the measure of the marquis de Châteaugiron of Rennes, which has the unqualified approval of all the vassals.'[23] A statistical analysis of these documents found that metrological standardisation is the fourteenth most common complaint of fifty mentioned, sandwiched between taxes and issues of 'personal liberties',[24] and repeated throughout the *Cahiers* are calls for 'one King, one law, one weight, and one measure'. As with Xun Xu's manipulations, measurement here is not only a practical matter, but symbolic of higher injustices and social order. The peasants were fed up with being shortchanged by their feudal masters and wanted their due.

———

The best example of an elastic unit from this period, and one that also embodies the realities of life and work, is perhaps the most surprising: the hour. For the majority of recorded history, the hour has had no fixed length, with the ancient Babylonians and Egyptians first dividing the day into twenty-four sections.[25] The ancient Egyptians counted ten units of daylight, adding two additional units for dusk and dawn, and marked units of night-time based on the movement of thirty-six constellations of stars called decans. Specialised scribes known as 'hour-watchers' looked for the appearance of a new decan over the horizon to mark the passage of the hours, while the nightly cast of twelve decans rotated every ten days, swapping in a new member as the Earth orbited the sun.[26] A total number of twelve may have been picked to match the lunar cycles of the year or perhaps the joints in the fingers of one hand, but charts recording these movements were important enough that they were inscribed on the interior lids of coffins. There they faced the unseeing eyes of the deceased to guide their souls across the night sky and reach resurrection with the dawn.[27]

However, with sunset and sunrise changing throughout the year, this meant the hour had to move with the seasons, expanding in summer and contracting in winter. These so-called temporal hours were inherited by Europeans in the Middle Ages and meant that the length of the hour in London, for example, could vary from 38 to 82 minutes.[28] This fact was not so much unnoticed in this period as much as it was an observation that would simply have made no sense. Thinking of the hour as a consistent measure was not a familiar concept for most people, while the minute and second didn't exist as common units. (The division of the hour into 60 minutes and the minute into 60 seconds comes from the Babylonians, who used a base-60, or sexagecimal, system of counting for their astronomy. The ancient Greeks later adopted this and divided circular astronomical maps into 360 divisions, which were later transposed on to clock faces.) The majority of peasants worked, instead, to the changing rhythms of the seasons, and smaller lengths of time were often accounted for using the same improvisational measures used for measuring long distances. A fourteenth-century cookbook, for example, instructs readers to boil an egg by advising that they leave it in the water 'for the length of time wherein you can say a *Miserere*.'[29] This is the 51st Psalm of the Bible and a plea for clemency that begins with the words 'Have mercy upon me, O God' – a nice bit of piety that I've found produces a rather runny egg.

The length of the hour was of particular interest to monks, who marked time to observe the system of seven canonical hours: points in the day when specific prayers are to be said.[30] These hours were *matins, prime, terce, sext, none, vespers,* and *compline,* with their timing communicated via church bells, the most reliable markers of time in the medieval world. They were not initially fixed either, as can be seen by the historical movement of the hour of noon across the day. Noon was originally *none,* and would be rung at around three o'clock in the afternoon, but slowly retreated backwards through the Middle Ages until it reached its current position of around midday in the fourteenth century. This was likely due to the fact that during periods of fasting, monks were not allowed to eat until *none.* This was a challenge during long summer days, prompting St Benedict in the sixth century to advise that *none* be said 'somewhat before the time, about the middle of the eighth hour'.[31] As historian Alfred Crosby notes, in the monks' system hours are treated as 'breadths, not points'[32] – a method of temporal ordering that is naturally variable. Although hours are inflexible today, we still make similar estimations in our own lives, telling ourselves we will keep working for the next few hours, or until lunch, while the length of unpleasant hours and the shortness of happy ones are experienced by everyone. It's one of the reasons I think people enjoy the night so much. For most of us, it is a temporal reprieve from the endless march of the clock, where hours regain their old elasticity like stiff sponges dropped in hot water, becoming more pliant and forgiving, and stretching out into the night. Until, that is, someone looks at a clock and, like Cinderella hearing those chimes at the ball, remembers that the regular schedules of life are bound to return tomorrow.

Temporal hours can be found in cultures around the world, and would only be sliced into regular portions by the tick of the mechanical clock (which the monks would have a hand in). Even then, a system of elastic hours has proved to have surprising longevity. In Judaism, for example, many observances that are expected to be performed at a particular time of day are based on relative hours. That means a ritual that takes place on the third hour of the day doesn't happen at three in the morning or three hours after sunrise, but a quarter of the way through whatever hours of sunlight there are on that particular day. In Japan, traditional timekeeping divided the day and the night into six units apiece, which varied in length with the seasons. This system survived the introduction of mechanical clocks to the country by Spanish missionaries in the 1500s, and was only abolished when the country adopted the Western calendar and time system in 1873.[33] This unusual coexistence of temporal hours and mechanical horology is memorialised by some of the most beautiful clocks in existence: circular timepieces with hour markers positioned on rails that move about their perimeter like tiny train carriages, sliding back and forth to adjust the length of each hour so that it matches the changes of the seasons.[34]

This acceptance of the primacy of nature in marking the hours of our days is also present still in civil contexts in the form of daylight saving. The practice of shifting the clock backwards and forwards one hour is maintained in most of the northern hemisphere, and though it's decried by many as an outdated concession to older methods of work, it's a reminder that the world doesn't always run on human time. Measurement may be the tool by which we impose structure on reality, but it still requires accommodation and concession in some domains.

The weighing of hearts and spirits

The pre-modern elasticity of measurement gives the practice a surprising sense of authority that can extend even into the realm of the mystical. Acts of measurement frequently appear in folklore, magic, and religion, often as a healing practice. Sometimes measurement is simply an authoritative ritual of care: an act associated with power and attention that gives the measurer some degree of control over their target. Other times it seems there is a belief in measurement's ability to capture some intangible dimension; a reflection, perhaps, of the lack of consistency in units. Such acts also show how the ordering quality of measurement could be adopted by other foundational aspects of society, including religion.

Perhaps the most widespread of these tropes is the weighing of the soul, a practice known in scholarship as *psychostasia*. Examples of this belief date back to the ancient Egyptians, who thought that the heart of the deceased would be compared to the weight of a single ostrich feather belonging to Ma'at, the goddess of truth and justice. If the supplicant had lived an honest and truthful life, then the feather would outweigh their heart and their soul would pass on to *Sekhet-Aaru*, or the Field of Reeds – a paradise of lush, endless

Measurement often takes on mystical significance, as illustrated by the ancient Egyptian belief in the weighing of the hearts of the dead

grasslands resembling the Nile Delta. If it tipped the scales, their bad deeds outweighing the feather's purity, then the heart would tumble on to the floor, to be eaten by the crocodile-headed goddess Ammit. This second death was final, which the Egyptians thought strict enough punishment for wrongdoers. This myth developed relatively late in ancient Egyptian history, only emerging fully during the New Kingdom period (around 1520 to 1075 BC).[35] Previously, evaluating the souls of the dead had been framed as something akin to a legal trial or courtroom, but it's thought that the change to measurement stemmed from a theological need to present this process as impartial.[36] With the weighing of the heart, there are no arguments to be made and no rhetoric to sway the minds and hearts of the gods, only the irreproachable judgement of the scales.

The metaphor of weighing one's fate appears in other cultures too: in the Greek *Iliad*, for example, Zeus consults a set of golden scales when the outcomes of certain duels hang in the balance. Here, judgement is pre-mortem and unconcerned with sin, but it's a matter of literary and theological dispute whether the scales decided the outcome of the fights or merely relayed a decision made elsewhere.[37] *Psychostasia* is thought to have made its way into Christianity via the Coptic sects of north-eastern Africa, who themselves likely inherited it from the Egyptians. The weighing of souls is never directly mentioned in the Bible, but Christian iconography has embraced the scales of judgement, usually assigning them to the archangel Michael, leader of God's armies.

The Bible itself illustrates how the practical matters of metrology have claimed authority in spiritual matters over time. In the five oldest books of the Bible, known as the Pentateuch or the Torah, advice on measurement is practical and didactic, a matter of social probity, like our earlier example of Talmudic law. Leviticus 19:35 in the King James Version states: 'Ye shall do no unrighteousness in judgment, in meteyard, in weight, or in measure. Just balances, just weights, a just ephah, and a just hin, shall ye have.' (*Ephahs* and *hins* being ancient Semitic units roughly equal to 15 kilograms and 6 litres respectively.) Later, Deuteronomy 25:13–14 states: 'Thou shalt not have in thy bag divers weights, a great and a small. Thou shalt not have in thine house divers measures, a great and a small.' Over time, though, these weights and balances are handled more symbolically in scripture, and when we reach the New Testament, written centuries after the Pentateuch, Jesus is using the notion of measurement purely metaphorically, telling attendees at the Sermon on the Mount: 'With what measure ye mete, it shall be measured to you again.' Measurement has become such a powerful symbol of justice that it can represent your moral deeds in life as well as your spiritual rewards and punishment. It's perhaps due to this symbolic potency that the Bible mentions measurement more often than it does charity.[38]

But the most intriguing examples of spiritual metrology are those rituals that offer practical help to those suffering in life. This is metrology as magic, often used for healing by measuring the body. In one story from a thirteenth-century collection of Christian miracles, a scholar is suffering from chronic pain. To cure himself, he takes a thread and first measures out his own height, then wraps the cord around parts of his body, calling out with every measurement the names of Jesus Christ and Saint Dominic, patron saint of astronomers. On reaching his knees, the scholar suddenly feels 'relieved of his pain' and exclaims, 'I am freed!'[39] In the same period, there was similarly a belief in the *mensura Christi*, or length of Christ. This appears in medieval manuscripts as an illustrated line, which readers were supposed to measure themselves, then multiply a certain number of times to discover the height of Jesus Christ. This was not just an act of devotion, but created a relic for the believer that was itself imbued with supernatural powers. 'If prolonged twelve times this segment shows the height [*mensura*] of the body of the Lord,' says one manuscript from 1293. Once the length has been calculated, another manuscript tells readers it should be kept close at hand: 'Those who wear this measure or keep it in their houses or see it every day cannot die of sudden death on that day . . . And they cannot be harmed by fire or water, nor by the devil, nor by a storm.'[40]

Such beliefs have survived longer than you might expect. North American folklore recorded well into the twentieth century offers similar cures. If someone is suffering from a wound or infection, for example, they are instructed to measure the area with twine and draw its outline on to the side of a living tree. This 'contaminated' surface should then be hollowed out and treated like the wound itself – daubed with ointment and covered in bandages – which allows the original to heal. And if a child is suffering from a chronic condition like croup or asthma, then they should be measured and a stick from a certain tree (often an elder) cut to match their height. This stick should then be placed somewhere significant: buried, often, or left in a corner of the invalid's room. When the child has outgrown the wood, they will have outgrown the disease also.[41]

Measurement, in these tales, is acting in different ways to draw on unseen powers. Measuring the height of Christ from the manuscript is an act akin to prayer for the faithful – 'a formula activated through recitation', as Emanuele Lugli puts it[42] – making the spiritual tangible by recreating Christ's body, which disappeared from Earth with the Ascension. For the pained scholar, measurement is a form of stocktaking and proof of attention, similar to the ministrations of a doctor. And with the folk cures from North America, measurement becomes a guarantee that the original and the duplicate are one and the same, allowing the sickness to transfer from the body to the object, leaving the former free and healthy. Such operations could not be carried out if measurement was restricted only to capturing the immanent; to measuring just what is in front of the ruler, so to speak. Like the elastic units of land, they reflect an understanding of measurement as an active intermediary, a tool that connects humans to the world around them and can be used to channel external forces. And it is an understanding of metrology that is not as dead as you might think.

Communal measures

As the Middle Ages progressed, new approaches to the standardisation and certification of units of measurement developed too, often matching changes in the political landscape. The city states of central northern Italy offer a fantastic example of this dynamic, demonstrating how the importance of communal maintenance of measures grew over this period. Thanks to a combination of fertile agricultural lands, trade routes to the East, and a burgeoning financial industry, this region of Italy was among the richest in Europe in the run-up to the Renaissance. Unlike the rest of the continent, which was controlled almost entirely by monarchies, it had far greater political variety, with many towns run as communes: polities that existed outside the system of feudal commitments. Over time, these city states developed their own governments, often with democratic characteristics, and their sovereignty included the right to set standards of weight and measure.

Such privileges were not always easily won. The period starts with this region under the rule of the Holy Roman Emperor, Frederick I, or Frederick Barbarossa (from the Italian *barba rossa*, or 'red beard'). Before assuming his throne and the Kingdom of Italy in 1155, he issued diplomas in which he claimed authority over measurement standards, alongside matters like controlling tolls on bridges and taxes in markets. In claiming this prerogative, he looked to his predecessor Charlemagne, the Frankish king who united much of central Europe through conquest in the eighth and ninth centuries and who oversaw numerous reforms, including metrological, creating an era of 'relative peace' that history enshrined as a golden age.[43] The difficulty, though, was enforcement. It's impossible to assess the impact of Charlemagne's reforms on day-to-day measurement, but they certainly maintained their power as an inspirational legend, one invoked by rulers like Barbarossa and by the masses when they felt they were being poorly treated. A thousand years later, on the eve of the French Revolution, the scientist Alexis-Jean-Pierre Paucton lamented that 'all measures were equal in the days of our earliest kings' but had been warped by the greed of feudal lords.[44] For Barbarossa, the right to set weights and measures was less royal prerogative than realpolitik: a precious commodity that could be traded during his attempts to impose authority on the independently minded Italian city states. In numerous peace treaties made through the twelfth century, he ceded authority over weights and measures to cities and local authorities, sometimes in perpetuity. These rights were celebrated by those who earned them, particularly by communities with strong democratic impulses.

In the republic of Siena in Tuscany, for example, weights and measures feature centrally in *The Allegory of Good and Bad Government*, a series of fresco panels created in 1338–9 in the commune's Palazzo Pubblico, or town hall. They are a remarkable artefact: rare secular paintings in an era of devotional art, and perhaps unrivalled in their didactic and allegorical ambition. These huge illustrations cover three of the walls in the *Sala dei Nove*, or Salon of Nine, where Siena's executive magistrates met to rule on city business, and any magistrate gazing about him during a long meeting would have been reminded of the power he wielded. On one side the fresco shows the effects of bad government, with farms burning in the countryside and violence in the cities. On the other side are the riches of good governance: farmers gathering abundant crops in their fields and transporting them to a bustling urban market where women dance in the streets. In the centre of the fresco is the huge figure of Justice, enthroned beneath the angel of Wisdom, who gently adjusts a set of scales. In the left pan is a figure representing Distributive Justice, who is busily beheading a criminal with one hand and crowning an honest man with the other. And in the right is Commutative Justice, who is handing down a trio of objects to a pair of waiting merchants. These objects are Siena's standards of measurement: the *staio*, or bushel, a unit of dry capacity; and the *passetto* and *canna*, two units of linear measure used for industry and construction.[45] Two woven cords lead down from these scales to be gathered in the hand of Concordia, or 'harmony', who holds them with the help of a group of Sienese citizenry.

It is a striking image that is both metaphorical and literal, didactic and utopian. Not only are standards of measurement given equal precedence to more traditional acts of justice (like beheading criminals), but merchants and citizens are shown as having an equal duty in their maintenance. Like the Sienese republic itself, reliable measurement required communal effort and cooperation. It bound people together, and in doing so ensured their mutual prosperity. Nearby on the wall, the text reads: 'Turn your eyes to behold her, you who are governing, who is portrayed here [Justice], crowned on account of her excellence, who always renders to everyone his due.'[46]

––––––––

For the benefit of standard units of measurement to be felt, though, there needed to be enforcement. In the thirteenth century, the Pisan mathematician Fibonacci noted that there were two types of merchant in the world: 'Those who measure with their arms, forearms, or steps' and those that 'use *pertiche* or another measurement standard'.[47] Working in the bustling docks and warehouses of Pisa, Fibonacci was in the perfect position to observe the metrological practices of the day, and wrote a number of treatises trying to educate his fellow citizens on these and other matters. In his 1202 text *Liber abaci*, he compiled a detailed catalogue of different units of measurement used throughout the ports of the Mediterranean – 'the *iugerum*, or the *aripennio*, or the *carruca*, the *tornatura*, or the *cultura*' – and noted that different units are applied to different materials, but that for fair measurement common units for all materials would be helpful.

Fibonacci's close attention to measurement reflects a 'new cultural awareness'[48] of metrology in this period, says Lugli, but treatises were not enough to bridge the gap between those who continued to measure with their body and those who aspired to greater precision. For that to be achieved, there needed to be public standards for consultation and officials to enforce them. These were not new innovations exactly, but seem to have fallen in and out of favour in Europe over the centuries, matching the rise and fall of more and less cohesive societies. We know, for example, that public standards were stored in the Acropolis in ancient Athens and on the Capitoline Hill in ancient Rome,[49] with copies distributed to markets and the like. We have examples of bronze and clay capacity measures found in the Athenian agora,[50] while in the preserved markets of ruined Pompeii you can see tables with inset basins representing different units of capacity. According to an inscription, the size of these units was regulated 'in accordance with a decree of the town councillors'.[51] It's difficult to say, though, how widespread these practices were, and excavations of ancient markets show that many unauthorised copies of weights and measures were also in circulation.[52]

In Italy, from the twelfth century onwards such public standards seem to have become popular again as the area grew in wealth. Lugli dubs these monuments *pietre di paragone*, or 'touchstones', as they were often carved into stone on the sides of important buildings and public infrastructure. This gives them a status similar to other important rights and duties in medieval communes, which were often recorded in the same way. The facade of the cathedral in Lucca, for example, tells traders to 'commit no theft nor trick nor falsification within the courtyard',[53] while the commune of Perugia announced a new form of taxation in 1234 by carving it into a slab of stone and affixing this to their cathedral's belfry.[54] The *pietre di paragone* got the same treatment, with units of length carved as stone incisions into the walls of churches and markets. Officials could then check their standards by inserting them into these spaces. Using these hollows to verify units was a simple and clever way to avoid tampering, as they could not be shortened and lengthening could be stopped with the use of metal end caps. As Lugli notes: 'While an object can be manipulated, a void is incorruptible.'[55]

Verifying the value of a standard this way is referred to by modern metrologists as 'traceability', and it underscores the notion that if units of measure are to be trusted, then there needs to be a way to trace them back to their source and ensure they have not been altered. While length standards were the most common example of *pietre di paragone*, other units also appeared carved in stone. A set of standards on Padua's Palazzo della Ragione from 1277 include a standard-sized brick, roof tile, and loaf of bread. The latter appears as a circle with a cross through the middle, like an oversized communion wafer. Paduans buying bread in the market could presumably hold up their purchase to the standard to check that it was the proper size. It is a symbol of good governance to rival that of the frescoes in Siena, with the local government literally certifying citizens' daily bread.

The use of public standards was commonplace for centuries, though by the time we move into the industrial era it's unclear how necessary they were. In London's Trafalgar Square, for example, a group of imperial length standards was fitted in bronze into the steps in the north-west corner in 1876. They show the lengths of yards, feet, links, chains, and rods (or perches), though it seems unlikely they were consulted after they were installed. A history of weights and measures by Surgeon-Lieutenant-Colonel Edward Nicholson, published in 1912, warns visitors: 'Anyone who has tried to get access to [the standards] in Trafalgar Square may regret that there seems to be no provision made against the site being made the usual lounge of often very objectionable persons.'[56] But such is the price of accessibility.

Along with these standards, inspectors were often deployed to check merchants' weights and measures against official units. This is a practice that, again, took place in the classical world, as with the ancient Greek *metronomoi*, who inspected weights and measures in Athens, and continued into the Middle Ages in various forms, from the *bullotai* of the Byzantine Empire to the *sensales* of Pisa. These individuals didn't just check measurements but could also offer an unbiased opinion on any product's quality, acting like free-roaming personal shoppers.[57] In England, measurement disputes in markets were settled by a special tribunal known as the 'court of piepowder' – 'the lowest and at the same time the most expeditious court of justice known to the law of England', as one legal history puts it.[58] Dating back to at least the eleventh century, the court of piepowder pre-dates the country's regular system of law courts and offered swift judgement on criminal activity that occurred on market day. Offences might have ranged from theft and assault to disputes over the proper measurement of goods. The court's name comes from a corruption of the French for 'dusty feet', *pieds poudrés*, referring to the soiled boots of travelling merchants, and this sort of mobile metrological justice still exists today. In the UK, for example, it is the job of Trading Standards Officers to investigate accusations of commercial fraud, checking that market scales have not been tampered with and pubs are pulling proper pints.

But perhaps the most pleasing example of the public confirmation of measures, and one that demonstrates how the certification of units can be the responsibility of a whole community, can be found in a sixteenth-century geometry textbook. The text is written by one Jacob Köbel, a town clerk and *Rechenmeister*, or 'reckoning master' (a teacher of arithmetic). The book offers readers many practical lessons, including how to measure the height of a tower from the ground and how to survey fields, but also shows how to replicate the rood, a long-lasting and widespread unit of length that found use all over Europe and that varied in size from 16.5 feet to 24 feet over the centuries. Rather than refer to any pre-existing standard or outside

This woodcut from a sixteenth-century geometry textbook shows a standard unit of measurement, the 'right and lawful rood', being constructed by the community

authority, Köbel advises that the budding metrologist can take matters into their own hands to define the rood. They should linger by the church door after the end of the service on Sunday and, when the congregation is exiting, 'bid sixteen men to stop, tall ones and short ones, as they happen to come out'. These men should be made to stand in a line, with 'their left feet one behind the other'. The length formed by this close-knit conga, says Köbel, is equal to 'the right and lawful rood'.[59] The resulting unit is one that is literally derived from the public body; built foot by foot from the community who will use it to measure the land they live on. This is measurement that ties together the lives of individuals to form a common understanding between them – a foundation of social order.

3

THE PROPER SUBJECT OF MEASUREMENT

_____

*How the scientific revolution expanded
measure's domain*

It would be an unsound fancy and self-contradictory to expect
that things which have never yet been done can be done except
by means which have never yet been tried.

—FRANCIS BACON, *NOVUM ORGANUM* (1620)[1]

Measure and the early modern mind

The early concerns of metrology spring from life's necessities: measuring soil to till and food to eat. But measurement as we understand
it today is equal to any phenomenon that can be named. We measure
rainfall and radiation; count calories, steps, and reps; and have units
capable of resolving both the depths of space and the emptiness of
atoms with equal precision. We even deign to measure things we
loudly declare to be unmeasurable, like happiness, pain, and fear. As
the poet W. H. Auden noted, we belong to those nations where 'the
study of that which can be weighed and measured is a consuming
love'.[2] But how did this love flourish? And when did measurement
assume such commanding authority over so many different aspects
of our lives?

If measurement was first embraced for its practical benefits, its rise
to epistemological prominence is due to its ability to tackle some
of the greatest puzzles of philosophy and science. This proving took
place during what is commonly referred to as the scientific revolution: a slow, chaotic, often contradictory upheaval of intellectual
culture that overtook Europe and beyond from the fifteenth century
onwards. What happened during this period has been framed in various ways: as a new centring of human subjectivity, as a 'secularization

of consciousness' that saw concern for the next world replaced with striving in this one, and as a material revolution of gears, lenses, and machines.[3] But whatever route we plot, the changes were undoubtedly immense. As the English poet John Donne lamented at the beginning of the seventeenth century, this 'new philosophy calls all in doubt / The element of fire is quite put out; / The sun is lost, and th' earth, and no man's wit / Can well direct him where to look for it.'[4]

Many intellectual shifts underpinned this revolution, including a scepticism towards traditional authority and a new appreciation for experimentation. But measurement and quantification were vital too. In the centuries preceding the early modern period (which begins around the end of the fifteenth century), intellectual convention placed limits on what was a worthy subject for measurement. Using scales to weigh grain was all well and good, but few thought that those same balance pans could help understand the nature of the cosmos or the mechanics of combustion. Indeed, there was well-established scepticism towards many empirical forms of knowledge, with theologians praising mathematics only insofar as it explicated God's creation, and natural philosophers (a group that precedes what we think of now as 'scientists') often embracing a vision of the world inherited from the ancient Greeks that explained the workings of natural phenomena primarily by reference to their underlying 'cause'.

In this understanding of reality, everything from acorns to water has a soul-like purpose, which is best understood through reason rather than experiment. As Aristotle wrote, 'men do not think they know a thing till they have grasped the "why" of it (which is to grasp its primary cause)'.[5] Measurement was the discipline of builders and traders, not thinkers. This was in part because the ancients and their followers considered that the sublunar world that we hear, see, and touch is one of constant flux; always becoming and never being, in Plato's formulation. Trying to understand it through observation was

like trying to study the night sky by watching its reflection on a shifting sea. When we rely on the 'instrumentality of the body for any
inquiry', warns Plato, our soul is 'drawn away by the body into the
realm of the variable, and loses its way and becomes confused and
dizzy'. Only the mind can reach beyond the confusions of the physical world to grasp something more permanent.

This attitude would slowly change throughout the Middle Ages,
with the shift best illustrated by a pair of Bible readings, written a
millennium apart but tackling the same passage from the deutero-
canonical Book of Wisdom. The verse in question states that when
God formed the world, he 'ordered all things in measure and number
and weight' – but what does this mean, exactly?

When Saint Augustine of Hippo, the most influential scholar of
late antiquity and father of the Western Church, unpacked these
words in the last years of the fourth century AD, he explained that
we should not imagine God counting and measuring the world,
as such concepts did not make sense prior to Creation. Instead,
'measure' refers to the divine principle that prescribes something's
limit; 'number' refers to the different forms that matter can take;
and 'weight' is the innate quality that draws physical objects to rest
and stability (all ideas taken straight from the ancient Greeks).
'Weight is not so much towards the bottom, but towards its place,'
Augustine notes. 'Fire tends upwards, stone downwards; by their
weight they are moved, they seek their places.'[6] Compare this to a
reading of the same verse in the fifteenth century by the German
philosopher and astronomer Nicholas of Cusa (1401–64) and we
see how these expectations will change. For Nicholas, measurement
is wielded by God just as humanity wields it for its lesser works, and
in his 1440 book *De docta ignorantia* ('On learned ignorance') he
explains that when we read that God 'ordered all things in measure
and number and weight', it means that 'in creating the world, God

used arithmetic, geometry, and likewise astronomy'.[7] Such practices
are the product of reason, he says, and by using them 'men surpass
beasts, for brutes cannot number, weigh, and measure'.[8] Nicholas
says that to carry out such practices is to take up the mantle of the
divine, for in measuring the world's mysteries we are following in
the footsteps of he who first 'balanced the fountains of waters and
weighed the foundation of the earth'.[9]

Qualities not quantities: the Greek inheritance

To better understand the revolution in measurement that took place
between the medieval and early modern periods, we need to understand the origins of what the American historian Alfred W. Crosby
calls medieval Europe's 'antimetrological' bias.[10] The ultimate source
for this antipathy is the ancient Greeks, whose works were trans-
mitted to Europe through translations and commentaries, often by
Arabic scholars, and synthesised into the dominant Christian culture by thinkers like Aquinas.

Plato was among the foremost of these influences, and one of his
most important doctrines was the division of the world into two
realms: the material realm and the realm of forms. The material realm
is reality as accessed by our senses – by touch, sight, smell, hearing,
and taste – while the realm of forms is eternal, immutable, and can
only be encountered by the mind. Forms include things that we refer
to today using abstract nouns, such as Beauty and Goodness, but also
the blueprints for everyday life, the adjectives that describe the world
around us, like Blue and Square and Cold. It's an approach to knowledge that stresses the world's intrinsic qualities – their essence, perhaps – and that seems to have permeated the culture. The Canadian
poet and classicist Anne Carson notes, for example, that ancient
Greek literature has a particular obsession with the inherent qualities

of people and objects. 'When Homer mentions blood, blood is *black*,' she writes. 'When women appear, women are *neat-ankled* or *glancing*. Poseidon always has *the blue eyebrows of Poseidon*. Gods' laughter is *unquenchable*. Human knees are *quick*. The sea is *unwearying*. Death is *bad*.'" These epithets are partly a result of the oral nature of early poetry, their repetition making the job of recitation easier, but also capture the nature of this philosophical tradition.

Forms could be embodied to a greater or lesser degree by entities in the material realm, says Plato. The sea and sky can reflect different degrees of Blue and Cold, while every circle you draw is an imperfect reflection of the perfect Circle, which can only be apprehended through rational thought, not the senses. This refraction of knowledge – the idea that the world as we experience it is somehow second-hand – is most famously illustrated in Plato's 'Allegory of the Cave'. In this story from *The Republic*, human existence is compared to a person chained inside a cavern, with a flickering fire behind them and restraints that stop them turning around. As objects pass in front of the fire, their shadows are projected on to the wall in front of the chained person. These shadows are all they can see – not the objects creating them. The objects represent the immaterial forms that comprise reality, while the shadows are humanity's perception of the world: imperfect and vague. It's an arresting image that still resonates today. There may not be many Platonists around now today (outside of philosophy departments, anyway), but we all feel, sometimes, like what we see and hear and touch is a shabby imitation of itself, somehow less than it should be.

For the ancient Greeks and those influenced by them, this metaphysics created a hierarchy of knowledge. It prioritises the transcendent over the material and the reasoned over the observed, in turn demoting measurement and quantification as tools for comprehension. In *The Republic*, Plato illustrates the fallibility of the senses with a simple demonstration. Hold your hand out in front of you palm up, he says, and look at your fingers. Examine your little finger, then your ring finger, then your middle finger and ask yourself, which is biggest? Your senses will tell you that your ring finger is both *larger* than your little finger and *smaller* than your middle finger, but how can this be? How can the same object be both larger and smaller? Plato's conclusion is that 'perception seems to yield no trustworthy result' and offers only a 'contradictory impression, presenting two opposite qualities with equal clearness'. This principle, he insists, is true not only of size, but of a range of qualities, from weight to firmness.[12] In such a world, measurement produces only inferior data. We can theorise about material reality, but its changing and contradictory nature undermines any conclusion we reach.[13] Tools of measurement have their use, of course, and Plato praises 'measures and instruments, and the remarkable exactness thus attained' when they are put to work in crafts like architecture.[14] But even the most pleasing building cannot match the solidity provided by philosophical thought. It is just another shadow on the wall.

Despite this attitude, the ancient Greeks and the medieval thinkers they influenced didn't dismiss mathematics as unimportant. Quite the opposite: they often revered number as a vital class of eternal forms, and saw certain mathematical pursuits (particularly geometry) as important demonstrations of transcendent knowledge. As long as they were only treated as abstract calculations and axioms, they were quite incorruptible. Writing in *De libero arbitrio* ('On Free Choice') at the end of the fourth century, Saint Augustine says: 'I do not know how long anything I touch by a bodily sense will persist, as, for instance, this sky and this land, and whatever other bodies I perceive in them. But seven and three are ten and not only now but always; nor have seven and three in any way at any time not been ten, nor will seven and three at any time not be ten.'[15]

But this perspective has more in common with numerology than mathematics as we think of it today. When Plato says that the ideal population for a state is 5,040 citizens, for example, he doesn't justify this choice by calculations involving, say, food supplies or division of labour. Having 5,040 citizens is right and proper for a city because 5,040 is the product of $7 \times 6 \times 5 \times 4 \times 3 \times 2 \times 1$ and therefore a mystically significant figure.[16] In this regard, Plato and his followers were influenced by the works of an earlier Greek philosopher, Pythagoras (c.570 to c.490 BC), who declared that 'all is number' – that numerical principles are a substratum of reality. The Pythagorean approach bestowed numbers with specific characters, so the number 1 was the monad, the singular God and the origin of all things, and 2 was the dyad, or the second cause and formation of matter. 1 and 2 were parents to 3, the first masculine number, which also happens to be perfect because it has a beginning, a middle, and an end. This is number as symbolism, not calculation, though historians argue it did at least present mathematics as a meaningful way to interrogate the nature of things.[17]

Augustine and his medieval successors inherited this symbolically lush understanding of number, and evidently thought it too interesting to abandon. They wove it into Christian beliefs, making numbers into 'the blueprint of a divinely-designed creation' and 'key to understanding the mind of Creation's God'.[18] So, the number 3 came to symbolise the Holy Trinity; doubling 3 produced 6, the number of days it took to create Heaven and Earth; and repeating it created 33, the presumed age of Christ at his death. The number 10, as the number of commandments, represented law, while 11, one greater than 10 and the number of disciples after the death of Judas, symbolised disorder and sin.[19]

Beliefs like this encouraged an attitude towards number that was qualitative, like the philosophy of the ancient Greeks. In the early medieval period, notes Crosby, numbers were deployed primarily for rhetorical effect. The hero of the eighth-century epic *The Song of Roland* warns his enemies that he will 'strike a thousand blows and follow them with seven hundred more', for example, while the Bible describes Israel defeating an army of 100,000 Syrian soldiers, the remnants of which are chased into a city, where 27,000 more are killed by collapsing walls.[20] Recipes might avoid precise units in favour of vague instructions like 'a bit more' and a 'medium-sized piece', while time and date, too, were treated with similar looseness.[21] This attitude was also influenced by the tools available. There were scales and measuring rods for practical tasks, but systems of number and calculation were limited in scope. For the majority of the Middle Ages, the default tool for reckoning with number was Roman numerals. With their combinations of tally strokes for digits under twenty and letters to represent larger numbers (V for 5, X for 10, L for 50, and so on), the system was easy to remember but ill-suited for calculation, with no mathematical symbols like the plus, minus, divide, or equal signs. It wasn't until the thirteenth century and the slow introduction of Hindu–Arabic that the concepts of place value and zero even became known.

The valuing of the qualitative over the quantitative in medieval thought can perhaps be seen most clearly in the work of the Latin encyclopedists – a diverse set of scholars that spanned countries and centuries, encompassing Romans like Pliny the Elder and Martianus Capella and European figures like the seventh-century Visigoth bishop Isidore of Seville and eighth-century English Benedictine monk the Venerable Bede. This group formed the backbone of European scholarship up until the twelfth and thirteenth centuries, drawing up what would eventually become the default medieval curriculum: the seven liberal arts. (These are the verbal *trivium* of grammar, logic, and rhetoric, and the mathematical *quadrivium* of music, arithmetic, geometry, and astronomy.)

They approached scholarship like indiscriminate magpies, compiling anything that caught their eye in disparate sources, and happy to treat rumour as reliable fact. Pliny's *Natural History*, for example, comprises thirty-seven books and includes knowledge cribbed from nearly 500 sources – an 'immense register [...] of the discoveries, the arts and the errors of mankind'.[22] Isidore of Seville's *Etymologies* is nearly as compendious, and takes, as its name suggests, the origin of words as the most reliable route to knowledge. Isidore explains that a word's history is shaped by the *causa*, or explanatory force, of the thing that it names, which often provides an answer to the basic question 'why?' – as in, 'Why does this thing exist and what is its purpose?' So the word *rex* ('king') comes from the monarch's obligation to act *recte* ('correctly'), while *homo* ('humans') are defined by our formation from *humus* ('the earth'). These etymological causes often incorporate a place, person, or behaviour associated with the subject, so the *vultur* ('vulture') is named after its *volatus tardus* ('slow flight'), while the mythical *basiliscus* ('basilisk') is so called because it behaves like a βασιλεύς ('king') to other snakes, scattering them before it. (Isidore here reassures us that despite the basilisk's fearsome reputation, it can easily be rebuffed with a weasel: 'Thus the Creator of nature sets forth nothing without a remedy.')[23] These etymologies take up the majority of Isidore's work, which was fantastically popular for many centuries and reproduced in hundreds of painstakingly handwritten manuscripts. Quantitative subjects like geometry, in comparison, are skimmed over quickly, while mathematics is praised mainly for its insight into biblical numerology. As Isidore writes, as if pre-empting readers' hatred of quantification: 'The reckoning of numbers ought not to be despised, for in many passages of sacred writings it elucidates how great a mystery they hold.'[24]

The information offered in a text like the *Etymologies* may seem to us trivial and haphazard, a collection of anecdotes that indicate only the past's primitivism. But just like the modern scientific method today, this qualitative understanding of the universe offered a sense of order in an otherwise chaotic existence. It is more than just an index of flora and fauna, but a cognitive map that helps orientate oneself in the world.

Aristotle and the Oxford Calculators

Plato's influence would continue to resonate throughout the Middle Ages, but from the thirteenth century onwards his position was gradually displaced by his student Aristotle, who had his own approach to observation and measurement in constructing knowledge. It's hard to overstate the centrality and importance of Aristotelian thought in this period, for although not everything he taught was accepted without question, the breadth and originality of his writing offered scholars a tantalisingly complete picture of the world. In an age when it was common to venerate the ancients, Aristotle's teachings were held in particularly high regard. Dante called him 'Master of them that know'; Aquinas praised him for attaining 'the highest possible level of human thought without benefit of the Christian faith'; while the Muslim philosopher Averroes summarised his contributions thus: 'The teaching of Aristotle is the supreme truth, because his mind was the final expression of the human mind.'[25]

Perhaps the most important aspect of Aristotelian thought was a shift in focus away from the abstract and towards the empirical. Aristotle still accepted Plato's theory of forms, but recast them as properties that inhered within matter, rather than concepts existing outside space and time. This relocation drew new attention to the material world and evidence of the senses. This was by no means the most important form of knowledge for Aristotle (in fact, it was only a stepping stone by which humans could reach higher, universal truths), but it meant that observations of what we see and hear could help construct knowledge, rather than distract from it. For Aristotle, the key to this was the process of induction, or proceeding from particular observations about the world to general theories. 'It is impossible to consider universals except through induction,' he wrote in his *Posterior Analytics*, 'and it is impossible to get an induction without perception.'[26]

As with Homer's *black* blood and *quick* knees, Aristotle understood that everything in the natural world possessed innate qualities that dictated its action. This is the aspect of his philosophy that most concerned the natural philosophers of the medieval period. They knew that while history tells us how things in the world *are*, their goal was to explain their *causes*. For Aristotle, the most important cause of things was their purpose, or *telos*. In Aristotle's teleological universe, smoke moves towards the heavens and rocks fall towards Earth for the same reason that rabbits run and wolves hunt. As the American historian of science Steven Shapin notes, it is a view that is close to animism: ascribing a soul-like purpose to inanimate objects. 'Aristotelian physics was in that sense modeled on biology and employed explanatory categories similar to those used to comprehend living things,' writes Shapin. 'Just as the acorn's development into the oak was the transformation of what was potential into what was actual, so the fall of an elevated stone was the actualization of its potential, the realization of its nature.'[27]

But even as Aristotle's teachings were disseminated across Europe, some of his methods began to be questioned. One of the earliest examples comes from the University of Oxford, where a group of scholars associated with Merton College in the early fourteenth century developed new mathematical and quantitative approaches to natural philosophy that have earned them the nickname among historians as the Oxford Calculators. This group of thinkers, which included Thomas Bradwardine, William Heytesbury, and Richard Swineshead, collectively produced treatises and texts that influenced European thought for centuries to come.[28] Bradwardine, the group's de facto founder, believed strongly in the application of mathematical principles to the natural world. It is mathematics, he wrote, that 'reveals every genuine truth, for it knows every hidden secret, and bears the key to every subtlety of letters.'[29]

The Calculators were particularly interested in Aristotle's explanations of motion and change, an expansive field of thought that captures phenomena we would not now think of as the domain of the sciences. To understand change, said Aristotle, was to understand nature, for nature was always in flux, morphing from one thing to another. Some of Aristotle's teachings here appear to us like scientific laws, such as his principle that nothing moves unless acted upon (a statement that Newton would later amend by noting that nothing *changes its motion* unless acted upon, to incorporate the principle of inertia). But other types of change are more metaphysical. Aristotle posited, for example, that all change is the product of four basic causes: the material cause (the object's composition), the formal cause (its arrangement of matter), the efficient cause (the action that brought about the object), and the final cause (its teleological purpose). In this schema, a statue's material cause might be some bronze or stone; its formal cause would be the statue's shape (say, a person or a horse); its efficient cause would be the sculptor who shaped it; and its final cause its aesthetic or religious appeal.

Here we can see how anachronistic it is to talk about a division of physics and philosophy in this period. The causes meant that for Aristotle and his followers, understanding motion involved both empirical observation and philosophical reasoning. The innovation of the Calculators was to lean much more on this first category than their predecessors. They developed Aristotle's thought with a new focus on movement as

Richard of Wallingford: abbot of St Albans, mathematician, inventor, and a contemporary of the Oxford Calculators

a phenomenon that could be observed within time and space, asking, when objects move through the world, how can we best describe (and measure) what we see? Their answer was to layer new distinctions on to the world. They disentangled old ideas, giving separate treatments of the *cause* of a motion (dynamics) and of its *effects* (kinematics); introduced new concepts like zero velocity; and created theorems to quantify these actions (like the Merton rule, used to calculate the average speed of a uniformly accelerating object).

All these gradations of meaning represent a key component in metrological thought. To measure something is to separate it from its neighbours; to say that this is most definitely seven – not six – and should in no way be mistaken for eight. It's a task that requires attention and focus, that asks us to stop assuming what we know and to try and create knowledge afresh. For the Calculators, this approach meant that they were able to overturn some of Aristotle's teachings, like the principle that the speed of an object was proportional to the force applied to it and inversely proportional to any resistance it encountered. Bradwardine noticed that this precept sounds reasonable, but doesn't actually describe reality. If it did, then any force could move any object – an ant could move a mountain – as long as it pushed for long enough. This isn't true, as objects can resist forces indefinitely if the resistance they offer is greater than the force applied. So Bradwardine amended Aristotle by claiming that in order to double the velocity of an object, you must square the ratio of force to resistance and so overcome the latter. To prove his theory, he even offered a general formula of sorts, which he wrote longhand, but which can be rendered in modern notation as $V \propto \log F/R$ (in which V is the velocity, F the force, and R the resistance).[30]

It's important to stress that the form of quantification that fascinated the Oxford Calculators does not map directly on to our understanding of the term today. The group did not carry out any actual experiments or use measuring tools like rulers or scales. Instead, they privileged abstract and verbal reasoning, and thought that their approach would let them measure not only motion, but anything that existed in gradations. That included things we think of as quantifiable today, like heat, light, and colour, but also more ethereal concepts, like grace, certitude, and charity (measuring the latter, they thought, might be a useful way to solve theological disputes). Nevertheless, their achievements were significant, representing an important transitionary moment in our understanding of measurement. Indeed, these insights would make Bradwardine famous enough to be mentioned by Chaucer alongside Augustine as one of the luminaries of the age, while one modern historian notes, somewhat tongue-in-cheek, that the Calculators' work earned Oxford's colleges a reputation that 'they never quite equalled before or after'.[31]

Measuring art, music, and time

The quantitative revolution of the early modern period didn't just take place in universities and in scholastic debates. Like ancient units of measurement, it was the product of craft too. Some historians have argued that the scientific revolution only took place at all because natural philosophers were influenced by the quantitative methods of practical trades. Writing in the 1940s, Austrian–American historian Edgar Zilsel said it was these 'artist-engineers' who influenced the first generations of scientists. Not only did they 'paint their pictures, cast their statues, and build their cathedrals, but also constructed lifting-gears, earthworks, canals and sluices, guns and fortresses, found new pigments, detected the geometrical laws of perspective, and invented new measuring tools for engineering and gunnery'.[32] It was these practical innovations, just like the temples and granaries of the ancient world, that proved the utility of measurement, allowing it to gain authority over new realms.

Consider one of the examples cited by Zilsel: the introduction of perspective in painting. Although ancient civilisations had some understanding of perspective in art, no culture had taken full advantage of its possibilities. The ancient Greeks made use of *skenographia*, or scene painting, in the theatre, for example – painting flat panels to create the illusion of depth – but limited its application to architectural facades like columns and porches. For most medieval artists, the rules of visual perspective were subordinate to more important hierarchies. Like the ancient Egyptians and Maya before them, they often made a scene's most important figure the largest in the frame, reflecting their temporal or spiritual power. They twisted perspective in Escherian fashion to highlight details that mattered most: showing the inside and outside of a building simultaneously,[33] or depicting events that happen at different points in time side by side. The motivation was to portray reality not as seen by the eye, but as understood in the mind.

This changed in the fifteenth century, thanks mostly to the work of a pair of literal Renaissance men: Filippo Brunelleschi and Leon Battista Alberti. The duo were inhabitants of Florence, then a wealthy and independent commune, and between them made important contributions to art, architecture, poetry, and philosophy. Brunelleschi seems to have been the first to devise and demonstrate pictorial perspective, transferring his architectural studies of Rome's ancient monuments into art through his use of the vanishing point (the spot at which parallel lines moving away from the viewer seem to converge). But he was a secretive man who never explained his methods. 'If you disclose too much about your inventions and achievements you give away the fruit of your genius,' he warned.[34] And it was Alberti who codified and disseminated the rules of linear perspective.

There are examples of perspective in painting prior to this, like the magnificent Ghent Altarpiece by the Van Eyck brothers, completed in 1432, but Alberti wrote the first theoretical primer on the subject, *De pictura*, in 1435–6, helping this doctrine spread through Europe.[35] In it, he advised readers using simple steps how to find the vanishing point for an image, and how, by sketching a foreshortened grid on to a canvas, they could use this visual scaffold to align and size its contents. To make the transfer of line from life to canvas more convenient, he advised that painters should erect a *velum*, or veil, between themselves and their subject:

> It is like this: a veil loosely woven of fine thread, dyed whatever colour you please, divided up by thicker threads into as many parallel square sections as you like, and stretched on a frame. I set this up between the eye and object to be represented, so that the visual pyramid passes through the loose weave of the veil.[36]

The 'visual pyramid' described by Alberti refers to medieval theories of optics. Prior to the thirteenth century, Western thinkers believed that vision worked via 'extramission', with the eye emitting rays that interacted with the world like a 'visual finger reaching out to palpate things'[37] (a mechanism captured by the Shakespearean imperative to 'see feelingly'). Thanks largely to the work of the eleventh-century Arabic scholar Ibn al-Haytham, known in the West as Alhacen, this was succeeded by an 'intromissionist' explanation, which reverses the causality so that it is the eye that receives impressions from reality. It's believed that these theories informed the work of artists like Alberti, encouraging the geometrical techniques of their perspective grids and creating a new incentive to divide the

In this illustration by Albrecht Dürer, a medieval draughtsman is using a quantified grid to capture the true perspective of his subject

world into spatially abstract units. After Alberti's primer was published, pictorial perspective spread rapidly across Europe, quickly supplanting older styles. One anecdote recounts how Paolo Uccello, a fifteenth-century painter whose geometric sketches are so precise and so detailed that they look like digital models, one night refused to come to bed and was overheard by his wife muttering under his breath: '*Oh, che dolce cosa è questa prospettiva!*' – 'Oh, what a sweet thing is this perspective!'[38]

A similar transformation took place in the world of medieval music, powered by another quantifying tool: musical notation. For most of the medieval era, music could only be passed on from singer to singer, with our favourite encyclopedist Isidore lamenting in the seventh century that 'unless sounds are held by the memory of man, they perish, because they cannot be written down.'[39] The dominant musical tradition of plainsong reflects these limitations. It consists of unaccompanied, monophonic chants, with the most well-known tradition, Gregorian chant, restricted to slow-moving melodies and a small range of pitches (often less than an octave). There are no dramatic changes in speed or volume, and as a result the music sounds pensive, even plodding, to modern ears. From about the ninth century, the first form of written music appears: the *neume* notation, represented by diacritic marks written above liturgical lyrics (the name probably comes from the Greek word for breath, from which we get modern words like 'pneumatic'). These indicated only relative pitch – whether a note was higher or lower than the previous one – with no sense of rhythm. An upward-tilted mark indicated the singer should go up on the next note, while a downward-sloping *grave* sent them down. These marks were initially recorded *in campo aperto*, or 'in the open field', meaning with no musical stave, but over time students and teachers began to trace horizontal lines across the page in order to make it easier to identify which *neumes* were higher and lower.

Fast forward to the eleventh century, and these lines have been stan-dardised as the musical stave. Skip forward again to the thirteenth century, and standardised note designs also appear, indicating how long to hold each sound.[40]

The ability to measure rhythm and pitch was expressed in poly-phonic music, where melodies are layered on top of one another in new harmonic intervals and rhythms move from pensive to playful. Whether notation was invented to capture increasingly complex mel-odies or the melodies themselves were inspired by the new notation it is impossible to say. But one was not possible without the other.[41] The new styles were known to contemporary theorists as *ars nova* and *musica mensurata* (the 'new art' and 'measured music' respec-tively), and as with any musical revolution, they were shocking to the establishment. The thirteenth-century Belgian music theorist Jacques de Liège complained that 'music was originally discreet, seemly, simple, masculine, and of good morals; have not the mod-erns rendered it lascivious beyond measure?'[42] while Liège's contem-porary, Pope John XXII, actually banned *ars nova* in the first papal bull to deal only with music, the 1324 *Docta sanctorum patrum*. Seven centuries later, the decree's complaint is quite cogent. It says that an obsession with melodic and rhythmic innovation has obscured the devotional content of plainsong: 'The voices move incessantly to and fro, intoxicating rather than soothing the ear, while the singers themselves try to convey the emotion of the music by their gestures.' The bull even refers specifically to the tools of musical measurement as responsible for this unwanted change, explaining that 'the mea-sured dividing of the *tempora* [periods, the basic unit of duration in music]' has allowed notes of 'small value' to proliferate, which choke and starve the 'modest rise and temperate descents of plainsong' like weeds in a well-ordered garden.[43] The skills of measurement, quanti-fication, and division were proving too fertile.

———

There's one other element of quantification in the Middle Ages, though, that overshadowed all others; that wrestled away the privileges of nature itself, and that has since beat its steady rhythm into all our lives. This was the invention of the mechanical clock, which began its global domination around the end of the thirteenth century. Europeans were not the first to design clocks using gears and mechanical linkages (Chinese and Arabian horologists beat them to it, building a number of impressively accurate astronomical clocks from the eleventh century onwards), but they introduced technical and theoretical refinements that propelled these systems into newly central positions in society.

Chief among the technical achievements were the oscillator and escapement: mechanisms that create the regular pulse we associate with all timekeeping today. Together, these act as a break for a suspended weight, the power source of medieval clocks until the coiled spring became common in the 1600s. This weight is hung from a cord in the middle of the clock and attached via a gear train to the escapement. Together with the oscillator, this geared wheel arrests the weight's downward passage, letting out its stored energy in small, incremental steps that become the regular tick of the clock.[44] Compared to the timekeeping systems of ancient civilisation, like the sundial and water clock, it offers a number of advantages. It works through the night, indoors or outdoors, and does not freeze in cold weather (important in the cold northern reaches of Europe). The integration of clockwork makes it portable (you cannot move a water clock without disrupting the flow) and also, in time, amenable to miniaturisation.

Perhaps most importantly, the mechanical clock propelled a new conception of time into the public consciousness, transforming it from a constant flow, embodied in steady emissions of water, sand, and mercury, to a quantified count; something divisible, discrete, and measurable. This count has only become more precise over the centuries, as improved technology has sliced the flow of time into ever thinner divisions. In the earliest clocks of the 1300s, the beat of an oscillator might last several seconds, but by the eighteenth century, the most advanced clock of its time, the marine chronometer, refined this to two beats per second. In the twentieth century, quartz crystals were introduced into digital watches, which vibrate tens of thousands of times a second. And the second itself is now defined by the vibrational frequency of an atom of caesium-133, which shudders back and forth an astounding 9,192,631,770 times to measure the duration of a single second.

Some of the earliest records of medieval clocks come from religious sources,[45] and it's thought that the demands of monastic life, with the monks' interest in regular prayers, spurred the invention of the mechanical clock.[46] Monasteries in this period were industrious dominions, with monks and servants toiling in libraries, mills, fields, and gardens, all while observing the Liturgy of the Hours. Bells were used to signal this schedule, and keeping such appointments was vital – not just to the order of the monastery but to the sanctity of one's soul. One eleventh-century French chronicler, Rodulfus Glaber, offers the cautionary tale of a demon who tries to convince a monk to ignore the call for matins in the night and sleep in: 'As for you, I wonder why you so scrupulously jump out of bed as soon as you hear the bell, when you could stay resting even unto the third bell.'[47] A similar concern can be found in the French nursery rhyme 'Frère Jacques':

Frère Jacques, Frère Jacques,
Dormez-vous? Dormez-vous?
Sonnez les matines! Sonnez les matines!
Ding, dang, dong. Ding, dang, dong.

Brother Jacques, Brother Jacques,
Are you sleeping? Are you sleeping?
Ring [the bells for] matins! Ring [the bells for] matins!
Ding, dang, dong. Ding, dang, dong.

The world of the monastery might have prompted an interest in 'rationalising and rationing' time, but it was the civic world that embraced the full potential of measuring it.[48] Clocks spread quickly through towns in the 1300s, with rich burghers often voluntarily introducing taxes to pay for their creation. They didn't have daily prayers to concern themselves with, but the daily routine of business: of opening markets and shops, of the arrival and departure of goods, of meetings and deals. The first clocks of this period were huge, expensive, and often lacked hands and faces. Instead, they announced the time via bells, just as in the monasteries, and the word 'clock' itself comes from the Old French *cloque* and Latin *clocca*, both meaning 'bell'. Just like measurement standards, these mechanisms were installed in public places – bell towers and city halls – where they could be seen by all. They were often incredibly complex, incorporating astronomical and zodiac calendars, as well as religious automata that moved about at set times each day. 'Something of the civic pride which earlier had expended itself in cathedral-building now was diverted to the construction of astronomical clocks of astounding intricacy and elaboration,' writes historian Lynn White. 'No European community felt able to hold up its head unless in its midst the planets wheeled in cycles and epicycles, while angels trumpeted, cocks crew, and apostles, kings and prophets marched and countermarched at the booming of the hours.'[49]

These devices were also viewed as public utilities that benefited citizens by encouraging industriousness. A petition to the Town Council of Lyon in 1481 says that the town 'sorely felt the need for a great clock whose strokes could be heard by all citizens',[50] while the inscription next to a clock installed in Caen in 1314 reads: 'I shall give the hours voice / To make the common folk rejoice.'[51] How the 'common folk' actually felt isn't clear, as the records we have from this period capture mostly the thoughts and feelings of the rich and powerful. But it seems that timekeeping inspired the same mixture of emotion it does today, with some welcoming their newfound sense of order and efficiency, while others resented the clock's restrictions and surveillance. As with the monastery bells, public clocks organised citizens into a cohesive unit, turning what had previously been private lives into communal tides that rose, worked, and retired as one. The fluctuations of the sun might determine the labour of agricultural peasants, but the new class of urban workers had previously lacked such central oversight and authority. This vacuum was filled. As Lewis Mumford noted: 'The clock is not merely a means of keeping track of the hours, but of synchronising the actions of men.'[52]

The clockwork universe

The clock was such a potent machine that it became the emblem of a whole new way of conceiving the universe: mechanical philosophy. This was the world view that most successfully opposed the wisdom of Aristotle, and that, by the end of the seventeenth century, would incorporate many elements of thought we now associate with the 'scientific' mindset, including the foregrounding of quantitative methods.

The mechanical philosophers reasoned that if clockwork was able to capture the movement of the stars and bring elaborate automata to life, who was to say that the natural world didn't operate under similar logic? Might not the universe itself be only a sort of monstrously complex clock, a *machina mundi*, or 'world machine', that animated matter through the operation of as yet undiscovered gears and levers? If this was the case, then Aristotle's teleological explanations, in which rocks fell to Earth and smoke rose to the heavens because it was in their nature to do so, were crude and unsatisfying. If nature worked like a machine, then it must rely on observable cause and effect, not some inscrutable soul-like purpose. You didn't need to grapple with the world of the forms to unravel its workings; you needed experimentation and observation. 'It is certain', wrote seventeenth-century French mathematician and philosopher René Descartes, 'that there are no rules in mechanics which do not hold good in physics [...] for it is not less natural for a clock, made of the requisite number of wheels, to indicate the hours, than for a tree which has sprung from this or that seed, to produce a particular fruit.'[53] In such a world, measurement mattered not just to weigh the fruit of the harvest, but to explain how it took root in the first place.

Thinkers like Descartes applied their mechanical rubric far and wide, devising explanations for phenomena ranging from digestion to magnetism (though Descartes himself drew the line at mechanising the mind, which led to his dualist outlook in which consciousness was the result of intangible operations). It was in astronomy, though, that mechanical philosophy would have its most important and lasting impact, transforming the animist world of Aristotle into an inanimate universe amenable to calculation. Since ancient times, humans have discovered in the regularities of the cosmos the impression of some greater order. But with the scientific revolution, the astronomer–priests, who advised kings based on their predictions of heavenly motion, would be displaced by the scientists, who offered fewer spiritual auguries but a clearer template to understand and control the natural world.

Until the middle of the sixteenth century, the cosmos was widely accepted to be geocentric, with the sun, moon, stars, and planets all orbiting the Earth, which was fixed immovably in a set of rotating concentric spheres. In Aristotle's telling, the universe was finitely large and unchanging (apart from its perpetual circular motion), with the heavens constructed from an indestructible form of matter not found on Earth. While the world we live in was comprised of four basic elements – earth, water, fire, and air – the world above embodied eternal perfection, and so was cast from an incorruptible fifth element known as *aether*, or the quintessence. This model sustained its authority for centuries, but close observation of the night sky using increasingly accurate telescopes revealed discrepancies. These were changes that belied its immutable status and movements that didn't fit the predictions of a simple geocentric universe. A lot of work was done to make the older models account for such eccentricities, but as they accrued mathematical caveats like sticky notes, doubts about their veracity became unavoidable.

The importance of the telescope supports the argument that scientific enquiry required the tools of craftsmen to advance. Telescopes were first developed by skilled lens grinders, who learned their trade making glasses, with surveyors and military officers providing some of the earliest customers. These two realms of skill, those of scholar and craftsman, were united in the figure of Italian astronomer Galileo Galilei, born in 1564, who made a living in his early adulthood designing compasses for artillery officers and whose work in physics, full of cannonballs and parabolas, seems to have been inspired by this field. It was Galileo who truly mathematised motion following the early attempts of the Oxford Calculators, using practical experiments to demonstrate flaws in Aristotelian wisdom. In one famous experiment, he dropped cannonballs and musket balls from the Leaning Tower of Pisa (an exercise that likely never took place in the way Galileo claimed) and showed that, contra to Aristotle, objects accelerate at a uniform rate, not proportionally to their mass. In other experiments he rolled balls down inclined planes, reasoning that if there were no impediment – if the ball were placed on a frictionless plane – then it might roll forever, an insight that undid Aristotle's principle that force was always necessary to maintain motion.

From 1609, Galileo's work moved to a new plane itself. Using home-made telescopes he'd constructed solely by reading descriptions of the device, he began to examine the night sky. There he found many unexpected sights, including the moons of Jupiter, the first objects seen orbiting other planets, and the surface of the moon (which, despite the claims of ancient authority, was not perfectly smooth but 'everywhere full of enormous swellings, deep chasms and sinuosities'[54]). Turning to the sun, he noted patterns of darkness moving from west to east across its face. Were these clouds? Other planets? Whatever they were, they showed the heavens weren't immutable, as Aristotle had claimed. The world above changed day by day, just like the world below. The mistake had been trusting tradition and authority instead of the testimony of the senses. As Galileo wrote, following his discovery of what we now know to be sunspots: 'So long as men were in fact obliged to call the Sun most pure and most lucid, no shadows or impurities whatever had been perceived in it; but now it shows itself to us as partly impure and spotty, why should we not call it spotted and not pure? For names and attributes must be accommodated to the essence of things, and not the essence to the names, since things come first and names afterwards.'[55]

Galileo had looked to the stars to find support for the theories of Nicolaus Copernicus, the sixteenth-century Polish astronomer who published his theory of a heliocentric universe in 1543. Copernicus's theories certainly upset the Church, but they didn't initially gain much traction in the intellectual world. Although his model decentred the Earth, and in so doing undermined cosmic assurances of humanity's special nature, it retained elements of Aristotelian systems, including uniform speed and circular orbits for the planets, and with these mechanisms their horrendous mathematical complexity. It would take the work of seventeenth-century German astronomer Johannes Kepler to simplify this system, using new observational data unavailable to Copernicus. Kepler gave the planets in his model elliptical orbits that sped up and slowed down as they moved closer to and further from the sun. These changes produced even more accurate predictions of planetary motion, but Kepler also saw his work as a direct challenge to the whole Aristotelian world view. In 1605, he wrote in a letter to a friend that he had renounced his belief that 'the motor cause' of the planets' motions 'was a soul' and was now 'much occupied with the investigation of physical causes'. 'My aim in this is to show that the machine of the universe is not similar to a divine animated being, but similar to a clock,' he wrote. 'This physical conception is to be presented through calculation and geometry.'[56]

This line of thinking, begun by Copernicus, Kepler, and Galileo, would bear its most influential fruit in the work of the British mathematician Isaac Newton. As a young boy, Newton was obsessed with making mechanical models out of wood, spending hours with his tools constructing and improving designs for windmills, sundials, and, most notably, a water clock. As a man, though, he would create the most complete and lasting picture of the mechanics of the universe yet devised with the publication in 1687 of *Philosophiæ Naturalis Principia Mathematica* (or *The Mathematical Principles of Natural Philosophy*). In it, Newton united the terrestrial and celestial realms through mathematics, devising his three laws of motion. The cannonballs of Galileo could stand in for the planets themselves, all operating under the aegis of Newton's universal law of gravitation. It was a totalising schema as ravenous as Aristotelian causes, able to accommodate every realm of the natural world in its calculus. No longer was there any distinction between the heavens and the Earth, and nor was there a clear sense of the universe's finite nature. The boundaries of the world had been flung open to infinity, its contents tracked through mechanics and mathematics.

In the wake of Newton, writes Steven Shapin, 'all natural processes were now conceived to take place on a fabric of abstract time and space, self-contained, and without reference to local and bounded human experience.'[57] This is a core dynamic in the history of measurement, in which greater abstraction allows our tools to encompass larger territories. Abstraction loosens ties to the local and particular, allowing freedom of movement. But it comes with a price. As we saw at the beginning of the chapter with Donne's concern that this 'new philosophy puts all in doubt', there is a psychological cost associated with the breaks of the scientific revolution, one that is powered by abstraction and measurement. The sureties of faith and the consolations of the ancients were weakened. New knowledge about humanity's literal and metaphysical place in the universe had to be assimilated. German sociologist Max Weber encapsulated this criticism as the 'disenchantment of the world': the means by which supernatural and mystic explanations of nature are replaced by the rational and scientific.[58] For some, the new existence is impoverished spiritually even as it's enriched materially. The displacement of Aristotelian physics, with its framework of animism and innate purpose, is part of this process. The world of Newton is one that's no longer ruled by qualities, familiar to us for their human feel, but by quantities, which work most efficiently when disconnected from the world.

And yet, this is not perhaps the full picture. Or, at least, any grand cultural changes we wish to acknowledge as a result of this work have to be let in gradually. Newton, perhaps more than any other figure from this period, embodies the rich contradictions of the age: the churn of belief and theory that to modern eyes seems so inconsistent. He was a rigorous and inventive experimentalist; someone who stared at the sun and poked blunt needles into his eye to investigate the nature of light. But he was also an ardent alchemist and numerologist who extended the logic of his theories far beyond what his experimental proofs could support (or so his critics maintained). And while his laws of gravitation and motion mechanised the universe, turning the music of the spheres into a cosmic game of billiards, he also stressed the importance of unseen forces in nature. He considered phenomena like magnetism and electricity beyond the scope of his philosophy, and placed immaterial principles at the heart of the *Principia* in the form of the gravitational constant, G, which somehow held the universe together. What exactly this force was Newton could neither quantify nor explain, but he knew it was necessary for his calculations to work. 'I have been unable to discover the cause of [. . .] gravity from phenomena,' he confessed, 'and I feign no hypotheses.'[59] Indeed, the idea of an invisible force that somehow permeates all physical matter, acting equally upon planets and people to what appear to be very different ends, does seem incredible. As the English economist and Newton scholar John Maynard Keynes put it: 'Newton was not the first of the age of reason. He was the last of the magicians.'[60]

4

THE QUANTIFYING SPIRIT

---

*The disenchantment of the world and the
history of hot and cold*

Some say the world will end in fire,
Some say in ice.
—ROBERT FROST, 'FIRE AND ICE'[1]

Dissecting the world

'And *this* was the great provocation,' says Anna-Zara Lindbom, slapping a sturdy wooden dissection table in the centre of Uppsala University's anatomical theatre in Sweden.

Lindbom and I are standing beneath the dome of a huge cupola built into the roof of the university's oldest building, the Gustavianum. Both theatre and cupola were built in the 1660s, positioned to take advantage of the natural light at a time when the work of dissection was moving out of the shadows of medieval taboo. It was here, on cold and crisp autumn mornings like today, that the bodies of executed criminals would be cut open for the benefit of young medical students. In return for their posthumous contributions to science, the dead earned themselves the redemption of a Christian burial, while students, watching from vertiginous tiers of wooden benches, learned how to save the living. Standing in the centre of the theatre, a chill in the air and the woodwork funnelling above my head, it feels like I'm on trial for some unknown crime.

'We're right across from the cathedral, you see,' says Anna-Zara, pointing out of the windows at the spires of Uppsala Cathedral, stabbing into the sky like stone spear tips. 'And that was the challenge from the university. For the students to come here' – she slaps the table – 'and dare to dissect what God' – she points at the window

– 'had created.' I nod and rub my hands together to generate a little heat, trying to imagine something other than the cold flesh of the dead.

As we're talking, a young man in a mustard blazer stamps up the stairs behind us. He walks into the theatre and by way of introduction dumps a body on to the table. It's plastic, thankfully: a life-size model with removable organs and a serene expression on its face. 'You'll have to finish up soon,' he tells us, as in just a few minutes a class of medical students will be heading our way for an introductory anatomy lecture. Although the Gustavianum is now a museum, the history and spectacle of the theatre are too good to ignore, and nearly four centuries after the cupola was constructed, Uppsala's students still gather under its light and learn how to dissect God's creation.

As we walk down the stairs, Anna-Zara, curator of scientific history at the Gustavianum, tells me about Sweden's 'obsession with measurement' – a period of devout quantification in the seventeenth and eighteenth centuries that was part of a Europe-wide drive towards the numbering of nature. As the achievements of figures like Kepler and Galileo penetrated the era's intellectual framework, it was clear that empirical techniques could deliver new insight and power. Historians refer to the 'quantifying spirit' of the age, an upheaval that swept through disciplines like an orderly poltergeist, replacing qualitative theories with those based on observation, experiment, and measurement.[2] 'They wanted to measure everything,' says Anna-Zara, as we walk through a room full of brass telescopes, balances, and surveying tools. 'They said: "We are going to measure the Earth, we are going to measure the skies, and yes, we will measure the people, too."'

This spirit was not merely intellectual, but political too: it was about dominion over the Earth. The English philosopher Francis Bacon argued in *Novum Organum* (1620) that it was the destiny of the human race to 'recover that right over nature which belongs to it by divine bequest'. Bacon said the mistake of intellectuals like the medieval scholastics had been to study words rather than things, which gave them power over logic but not nature. He advocated instead for a method of empirical enquiry based on 'tables of discovery': systematic records of material phenomena that could be used to reconfigure the world to man's advantage. In his utopian fable *New Atlantis*, Bacon sketched out a blueprint for his ideal state: an island empire ruled by pious monarchs, whose power is sustained by what Bacon calls 'Solomon's House', a state-run scientific organisation that is equal parts secret society and research university. The institute's credo encapsulates the grand ambitions of these early empiricists: 'The end of our Foundation is the knowledge of Causes, and secret motions of things, and the enlarging of the bounds of Human Empire, to the effecting of all things possible.'[3]

It's not surprising, then, that Sweden's own mania for measurement coincided with the country's rise to great power status and a period of military expansion. Like many European nations, the bureaucracy of the Swedish state grew in complexity and power during the early modern period, marshalling the country's resources and channelling them into trade, industry, and military might. Swedish troops, drilled with precision and bearing the latest armaments, had a reputation as some of the best in Europe,[4] conquering swathes of modern-day Finland, Norway, Russia, and Estonia. Sweden's invasion of the Polish–Lithuanian Commonwealth in 1655 was so swift and brutal, leaving a third of the dual state's population dead, that it's known to historians simply as 'the Deluge'.[5]

The link between measurement and conquest is made plain as Anna-Zara stops to point out an illustration of Olof Rudbeck, the seventeenth-century professor of medicine who oversaw the creation of Uppsala University's anatomical theatre. In his picture in the Gustavianum, Rudbeck is wielding a scalpel, but rather than dissecting a body he's cutting into a miniature Earth, peeling back the top layers of the planet like an orange. His claim to fame was tracing the flow and discharge of the body's lymphatic system, which provides essential drainage and helps combat infections; but he also pioneered a new discipline: the archaeological study of rock strata, now called stratigraphy. It was Rudbeck's belief that Sweden was the site of the lost civilisation of Atlantis, and that modern Swedes were descendants of this ancient race; and after his success mapping the lymphatic system, he drew his scalpel across the land, mapping the clay beds and eroded channels of Sweden, which he said were the remainders of the lost canals and waterways of Atlantis. It was a country-wide dissection he hoped would reveal the very organs of Sweden's greatness.

At the Gustavianum, though, I'm here to see a particularly notable product of Sweden's measurement craze; a metrological artefact that defines a phenomenon we encounter every day of our life. It's the first ever Celsius thermometer, hand-labelled by Anders Celsius himself (1701–44), a former professor of astronomy at the university.

Anna-Zara guides us to the case containing Celsius's thermometer, but it still takes me a moment to locate it. It's an artefact that stands out among the museum's metrological tools not for its complexity, but for its delicate, almost fragile construction. It sits in a wooden frame, fastened to a faded piece of paper on which are written two separate temperature scales. The instrument is made from glass but filled with silvery mercury, giving it a mirror finish. The stem is thin, just two millimetres in diameter, while the bulb looks perilously swollen, like a ripe silver pear. It was with this instrument on Christmas Day in 1741 that Celsius took his first temperature readings in the snow,[6] before going on to establish the now familiar temperature scale that bears his name, with 0°C as the freezing point of water and 100°C as its boiling point. As Anna-Zara explains, Celsius's contributions to the field of thermometry didn't consist of grand theoretical achievements. He didn't even build his famous thermometer – that was the work of the French astronomer and instrument-maker Joseph-Nicolas Delisle, inventor of his own rival temperature scale. 'Celsius's innovation', says Anna-Zara, 'was making his scale reliable. He was a great fan of testing and experimentation. He did it everywhere he went. He wanted to *prove* everything.'

Establishing such reliability is a cornerstone of measurement, and was incredibly challenging in early thermometry. Today, the only confusion most of us face regarding temperature is when we're confronted with Celsius when we're used to Fahrenheit, and vice versa. (As a unit of the metric system, Celsius is the global standard for temperature, while Fahrenheit is used in the United States and a handful of other nations.) But when Celsius himself was investigating the problem of hot and cold, there were at least thirty competing temperature scales in Europe, each of uncertain reliability. The problem was that the basics of thermometry were well known enough for anyone to make their own temperature scale, but no one knew how to make devices that were consistent and reliable.

Hasok Chang, a professor of history and the philosophy of science at the University of Cambridge, has written about these problems extensively. The creation of reliable thermometers, he says, perfectly demonstrates the difficulty of establishing scientific truths in unknown territory. In the case of temperature, the problems are simple to articulate but maddeningly difficult to answer. How do you test the reliability of a thermometer without already possessing a reliable thermometer as a benchmark? Thanks to the work of Celsius and other scientists, we can say now that water freezes at 0°C and boils at 100°C, but how would you establish those facts *before* you have trustworthy thermometers with which to make your measurements?[7] Such problems seem like the epistemological equivalent of hammering a nail into thin air, asking us to conjure up a certain point of contact with the world in the absence of a reliable framework of knowledge.

This dilemma occurs throughout the history of metrology and demonstrates both the discipline's challenges and its utility. In our attempt to measure uncertainties we construct new knowledge, and in doing so we inevitably remake our understanding of the world. Celsius's own thermometer offers a simple illustration of this fact. Look a little closer at its faded paper scales and you'll see that despite our understanding today of temperature as something that *rises* with the heat, Celsius himself thought the opposite. His original scale actually places 100°C at the bottom for the temperature at which water freezes and 0°C at the top for water's boiling point. One suggested explanation for this is environmental. In Sweden, temperatures are frequently colder than freezing, so ordering the scale from 100 to 0 would have let Celsius easily record temperatures in winter without resorting to negative numbers. It wouldn't be inverted until years later, though even today we're not quite sure who's responsible. Few things better illustrate the constructed nature of our metrological knowledge than a world where water boils at 0°C.

'A fountain that drips in the sun'

The measurement of temperature might seem inconsequential compared to other phenomena. Heat and cold are not integral to activities like trade and construction, nor do they dominate our conceptual understanding of the world, as with the measure of time. But this attitude is a habit of modernity, the result of our domestication of temperature. Thanks to air conditioning and central heating, down jackets and hand-held fans, humans are able to blunt extremes of cold and heat. Fire and ice still exist as threats but are mostly controlled in our daily lives. Heat is something to be adjusted with a thermostat; cold to be locked away in ice cubes and freezers. For most people living in the developed world, it's only really through the effects of climate change, when the planet throws wildfires, hurricanes, and blizzards at us, that we've begun to remember how difficult and unruly temperature can be.

This understanding was certainly closer to the surface for our ancestors. Temperature as a measurable concept didn't exist in the past, but heat and cold were recognised as animating principles of the natural world. For the ancient Greek philosopher Heraclitus, fire was not merely a material phenomenon but the first principle of the universe. It was the source of all life and a constant roiling change that burned through the world, transforming matter. The order of things 'no god nor man did create', taught Heraclitus, only 'ever-living fire', which shapes life in the womb and burns dead wood to make space for new growth. 'All things are an exchange for fire, and fire for all things, as goods for gold and gold for goods,' he said.[8]

Some centuries after Heraclitus, Plato and Aristotle retained the importance of heat in their philosophies, though complicated the picture by making it one of four contrary qualities: hot and cold, and wet and dry. Later thinkers like Hippocrates in the fifth century BC and Galen in the second century AD developed this premise further by incorporating heat into the doctrine of the humours, a long-lived system of medicine and morality that dominated Western understanding of the body for more than a millennium. Heat, for Galen, was the fuel that powered life's journey. It moulded organisms in the womb and propelled growing children, before leaking out of the body with old age. It could be generated anew in adults through exercise and the consumption of certain foods (wine was a favourite),[9] but when it had left the body, death was not far behind. Such an understanding of the world is both mystical and intuitive. We don't need to see a body on a dissecting table to know that heat is life and cold is death.

The fundamental position of temperature naturally led to further scrutiny. Galen was one of the first thinkers to suggest there might be 'degrees of heat and cold' and attempt to distinguish between them.[10] (He thought four degrees was probably all that was needed.) These degrees weren't measured quantities, though, but shorthand labels for vaguely defined categories. Four degrees of heat was a lot, three degrees was less, and so on. Around this same period, people were creating instruments that registered such temperature changes. These usually consisted of a container of water with a hollow tube poking into its interior like a straw. As the temperature changed, air in the vessel contracted or expanded, creating a vacuum that pulled water up the tube – the same principle used in air thermometers. The first-century Roman mathematician and engineer Hero of Alexandria is one of the first to describe such a device, which he called 'a fountain that drips in the sun.'[11]

It was not until the 1500s, though, that natural philosophers began to add numbers to these instruments, with the Venetian physician Santorio Santorio and our stargazing Galileo Galilei among the front-runners. Santorio, like Galileo, was a consummate experimenter, though as a physician he applied measurement to the body with particular zeal, inventing the pulse-measuring *pulsilogium* ('the first instrument of precision in the history of medicine'[12]) and constructing a human-sized 'weighing chair' to better understand human metabolism. Santorio himself sat in this giant pair of scales while he worked, slept, ate, drank, defecated, and urinated, weighing himself before and after each activity in order to calculate the exact input and output of matter. His aim was to balance these factors and so reach 'a perfect Standard of Health as will last to a Hundred Years.'[13]

We know that both Santorio and Galileo built devices similar to Hero's fountain but deployed them in a more rigorous fashion, using them to compare the heat of specific objects and environments. These were what we should call thermoscopes rather than thermometers, as they lacked proper scales. The scales on these devices were ordinal, meaning they were arranged in a hierarchy from top to bottom, but they were not interval scales, where each degree is uniform. Think of the difference like this: a customer survey that asks you to rate your happiness from one to five is an ordinal scale because you are ranking your response. But it's not an interval scale because the exact levels of happiness are subjective and vaguely defined; they cannot be added or subtracted from one another. There's no way to know if your five isn't your neighbour's three.

Santorio directed his thermoscope at the human body, publishing a text in 1626 describing an air thermometer marked with degrees, with one end placed in the human mouth. For the most engaging

Before reliable thermometers were created, scientists like
Venetian physician Santorio Santorio used thermoscopes
like this to roughly estimate temperature – in this case, a patient's

records of Galileo's devices, we have the letters of his friend Giovan Francesco Sagredo, who built several thermoscopes based on the astronomer's design and reported his own findings. At the height of summer, Sagredo notes with good humour in a letter to Galileo that all his interests lie in 'measuring the aforesaid heat and cooling the wine', while in the colder months he reports discovering 'various marvelous' phenomena, including the observation that the air is often colder than snow, that snow mixed with salt is colder still, 'and similar subtle matters'.[14]

These experiments show how quantification can transform a concept like temperature. With the help of these early thermoscopes, hot and cold are no longer qualities that inhere within objects, hidden and inscrutable, but information that can be extracted from its source. Transformed into abstract data, this information can be collected, shared, compared. And the new testimony of these instruments is so powerful that it can overrule even our senses. Sagredo notes that the water in natural springs is colder in winter than in summer, 'although our feelings seem to indicate the contrary'.[15] And writing at around the same time in 1620, Francis Bacon comments that the thermoscope's sense of 'heat and cold is so delicate and exquisite, that it far exceeds the human touch'.[16] The scientific instrument had begun to displace human experience as the arbiter of reality.

In 1624, the word 'thermometer' was coined by the French Jesuit priest Jean Leurechon, derived from the Greek words *thermos* ('hot') and *metron* ('measure').[17] At this point the device had taken on a familiar design: 'an instrument of glass which has a little bulb above and a long neck below' that rests in a reservoir filled with 'vinegar, wine, or reddened water', as Leurechon puts it. Moving the instrument from a cold to a hot place causes the liquid to fall as the air 'rarefies and dilates and wishes to have more room, and therefore presses on the water', he says, while moving from hot to cold has the opposite effect as the 'air is cooled and condenses'. This phenomenon is simple and reliable, even if the underlying causes were unknown, and Leurechon writes almost wistfully of this magic, noting that you can sway a thermometer by breathing on it 'as if one wished to speak a word into its ear to command the water to descend'.[18]

Other instrument-makers took this supernatural element to extremes, including the Dutch tinkerer Cornelis Drebbel, a contemporary of Galileo who was one of the most famous inventors of his day. An eighteenth-century historian describes him as a 'kloeck verstant, een pronck der Wereldt' – 'a bold mind, a show-off to the World'[19] – while those who met Drebbel talk of a gentle but restless figure whose inventions ranged from the practical (a lens-grinding machine) to the fantastical (a harpsichord powered by solar energy). At the court of King James I of England (James VI of Scotland), Drebbel earned his keep as in-house showman, there demonstrating his most famous invention: the *perpetuum mobile*, or perpetual motion machine. Details of the device's construction are hazy, but it seems to have combined elements of a traditional astrolabe with a perfectly circular tube full of ever-moving liquid.[20] Drebbel said the *perpetuum mobile* was proof of his alchemical achievement and mastery over arcane forces, including 'the fiery spirit of the air'.[21] But contemporary descriptions of this instrument, including an extremely sceptical letter written to Galileo by a former student, suggest it was actually an early thermoscope, and that the restless liquid that sloshed around its interior was animated by nothing more ineffable than the weather. Drebbel's machine is an oddity, but it marks a transitional moment in intellectual thought: a period when a rudimentary scientific instrument can be presented and accepted as a mystical device.[22]

Fixing the points

The explanatory potential of thermometers shows why scientists were so keen to standardise the measurement of temperature. Although the exact mechanisms of heat and cold were not fully understood in the early modern period, scientists knew that changing temperature had an effect on a range of phenomena, from subtle processes like the speed of chemical reactions to more obvious events like melting, evaporation, and condensation. This meant that being able to record and subsequently adjust temperature was essential for experimentation in a range of disciplines. If thermometers were to be useful in this work, then their scales had to be consistent and verifiable. There had to be a common language of heat and cold.

By the seventeenth century, a number of early thermoscopes had gained scales divided into degrees, but these markings were arbitrary and particular to each instrument. Writing in 1693, the English astronomer Edmond Halley, discoverer of the eponymous comet, complained that every thermometer works 'by Standards kept by each particular Workman, without any agreement or reference to one another'. The consequence is that 'whatsoever Observations any curious Persons may make ... cannot be understood, unless by those who have by them *Thermometers* of the same Make and Adjustment'.[23]

Creating consistent temperature scales required the discovery of certain stable thermometric markers: phenomena that were consistent in temperature whenever and wherever they appeared. In the same way that the consistency of millet seeds provided a baseline for creating ancient measures of weight and length, allowing anyone to recreate the same unit as long as they could find enough seeds, scientists needed fixed points to anchor their scales. Early suggestions seem, in hindsight, maddeningly imprecise, and included phenomena like the hottest day in summer, the heat of a single candle, and the cold of certain cellars in Paris.[24]

One of the earliest calibrated scales of this sort was devised by Isaac Newton, who, in 1701, created a thermometer using linseed oil, with degrees defined using a number of would-be fixed points. These included the temperature of air in winter, spring, and summer; two separate points based on 'the greatest heat of a bath which a man can bear' (in the colder one he moves his hand constantly, in the hotter the hand is kept still); and other benchmarks based on 'the external heat of the body in its natural state' and 'that of blood newly drawn.'[25] These phenomena may seem as thermometrically vague as melting butter, but some are surprisingly reliable. Body heat, for example, is generally stable, having been fixed in place by millions of years of evolution. Our bodies have been tuned over time to perform optimally in a narrow band of temperature and regulate this through a host of physical reactions, including sweating, shivering, and the dilation of blood vessels. As a result, the internal temperature of most bodies, most of the time, is 37°C (or 98.6°F), with even a few degrees of variation causing serious harm.

By the turn of the eighteenth century, two innovations were beginning to move thermometry into the realm of the reliable. The first was the slow-gathering consensus that the freezing and boiling points of water provided the most convenient and consistent thermometric benchmarks. The second was a series of technical advances that helped to establish the reliability of these points. These included the spread of sealed thermometers and the use of different liquids as the medium within. The paradox of thermometry remained, though: how do you construct a reliable thermometer from fixed temperature points without already possessing a reliable thermometer to confirm that these points are fixed? The solution to this quandary would take decades of patient work from dozens of individuals.

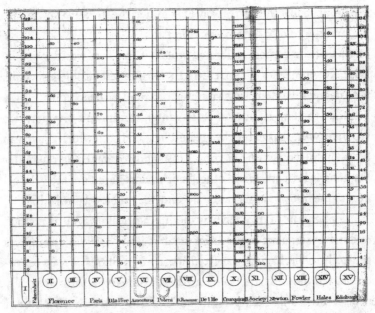

*In the 1700s, there were numerous competing measures, each
with their own fixed points of thermometric reference*

One contributor who stands out is Daniel Gabriel Fahrenheit, an instrument-maker whose work brought him fame in the early eighteenth century, but whose early life was marked by tragedy. Born into a wealthy family in Danzig (modern-day Gdańsk), Fahrenheit was orphaned at the age of fifteen when his parents died after eating poisonous mushrooms. His legal guardians arranged for him to be apprenticed in a trading business in Amsterdam, but Fahrenheit found the world of bookkeeping tedious. Instead, he yearned for the scientific pursuits he'd enjoyed during his early schooling. After four years in Amsterdam, he absconded from his apprenticeship and became a scientific fugitive, stealing money from his employers to fund his own research, while hopping around European cities to learn from the great scientists of the age. His guardians responded as any caring adults would: they had a warrant issued for his arrest and gave the authorities permission to deport him to the East Indies if captured.[26]

The warrant, thankfully, never caught up with Fahrenheit, and from his twenties he found himself drawn into the world of scientific-instrument-making, a vocation that involved both theoretical knowledge and practical proficiency. He began specialising in glass blowing, building thermometers, altimeters, and barometers and earning a reputation as, in the words of one contemporary, 'that industrious and incomparable Artist, Daniel Gabriel Fahrenheit'.[27] In 1708, he met up with Ole Rømer, a famed Danish astronomer who also happened to be mayor of Copenhagen. (It's possible Fahrenheit sought him out to try to clear his then still outstanding arrest warrant.) Rømer had devised his own temperature scale, using the sexagesimal numbering system that was familiar to him as an astronomer. He had set the top of his scale, the boiling point of water, at 60 degrees, and the bottom, the freezing point of brine, at zero. Fahrenheit decided to adapt Rømer's scale with a few key changes.

Firstly, Fahrenheit was dissatisfied with the 'inconvenient and inelegant' fractions that marked freezing point and body temperature in Rømer's scale (7.5° and 22.5° respectively). He bumped these up to 8 and 24 to make them neater, and then multiplied the whole scale by four, creating smaller intervals between degrees and thus finer accuracy in readings. These changes give us the temperature benchmarks familiar to Fahrenheit users: 32°F for freezing water and 96°F for body temperature (though this fixed point is 2.6°F below current estimates).

Fahrenheit also took the step of using mercury instead of the then common 'spirit of wine' (ethyl alcohol) as the measuring medium inside his thermometers. Mercury, or quicksilver as it's long been known, not only has a higher boiling point than alcohol, allowing the thermometers to be used in a greater range of temperatures, but it reacts more quickly to changes in hot and cold and expands and contracts to a greater degree. This meant thermometers could be built on a smaller scale, while still registering the same range of temperatures. As is often the case in metrology, it was not any one change that made Fahrenheit's instruments stand out, but his ability to bring together a range of improvements. His thermometers were so renowned he was inducted into Britain's Royal Society in 1724 for his work, with his temperature scale adopted by English-speaking countries around the world until it was displaced by its metric rival.

———

Fahrenheit's practical genius rested on foundations created by others. Specifically, it was trust in the fixity of the boiling and freezing points of water that anchored each end of his scale like tent pegs holding down a guy rope in high winds. Much of this confidence was due to the work of Anders Celsius, who pioneered the now familiar centigrade scale, with its 100-degree range between the freezing and boiling points of water, yet even Celsius's work was incomplete.

Think about what constitutes the 'boiling point' of water in his scale. Not only is the temperature at which water boils affected by criteria like water purity, atmospheric pressure, and the depth of the vessel used, but boiling itself is a term of not inconsiderable imprecision. Does water boil when the first bubbles appear, or when they are produced in a continuous stream? A temperature scale from the 1750s built by one of Britain's most respected instrument-makers, George Adams, demonstrates this problem clearly: it has two markings for separate boiling points, one at 204°F, where 'water begins to boyle', and one at 212°F, where 'water boyles vehemently'.[28]

Nailing down this end of the scale became a priority for many scientists in the mid-to-late eighteenth century, with the Royal Society even appointing a special seven-member task force in 1776 to solve the problem.[29] The work required a fantastical level of close attention, with scientists exhausting every configuration of boiling water. The research of one member of the committee, Swiss geologist and meteorologist Jean-André de Luc (1727–1817), stands out for its dedication and attention to detail.

In one series of experiments, de Luc explored the realm of superheating, where the temperature of water is raised above 100°C without boiling. He found that removing oxygen from water assisted this process, but without access to any mechanical equipment that could deoxygenate samples for him, he had to manually shake vessels by hand to remove the gas, like someone shaking the bubbles from a fizzy drink. 'This operation lasted four weeks,' he writes in one report, 'during which I hardly ever put down my flask, except to sleep, to do business in town, and to do things that required both hands. I ate, I read, I wrote, I saw my friends, I took my walks, all the while shaking my water.'

Elsewhere, in a 1772 treatise, de Luc sets out to determine exactly what constitutes *la vraie ébullition,* or true boiling, and instead finds only a multitude of phenomena forced into homogeneity by this single, restrictive term. He examined ebullition by watching pots of boiling water with the attentiveness of a new parent leaning over the crib, noting the speed, size, and sound of bubble formation; at what depth they appear in the water; whether they make it to the surface or implode mid-rise; and the level of agitation on the water's surface. He distinguishes between numerous new categories, including water that is *sifflement* ('whistling'), *bouillonnement* ('bubbling'), and *soubresaut* (a ballet term for a quick, short, vertical jump, referring here to the movement of the bubbles). Although measurement is stereotyped as a stultifying activity that reduces the vibrancy of the world to mere numbers, work like de Luc's shows the opposite can be just as true. The desire to measure something with accuracy forces people to seek new corners of the phenomenological world; to find nooks and crannies of physical experience that were previously lost in the melee. The closer we look, the more the world reveals itself.

The irony of the investigations by de Luc and his peers is that they revealed that the boiling point of water was not the reliable thermometric marker many had considered it to be. There was just too much variety in the temperatures at which boiling occurred. Instead, scientists turned to measuring the steam produced by the water, which proved to be a much more stable reference point. Whether the water below was *sifflement* or *soubresaut*, the steam above was consistent. This may seem like a failure, as if the scientists working on the problem had wasted time with a series of wrong answers before stumbling upon the 'right' solution. But it demonstrates an important concept within metrology and science, what Chang calls 'epistemic iteration'. This is the process by which 'successive stages of knowledge, each building on the preceding one, are created in order to enhance the achievement of certain epistemic goals.'[30]

In the case of thermometry, this process starts with human experience of hot and cold, which is then replicated with the first basic thermoscopes. They confirm obvious truths: that snow is colder than fire and summer is warmer than spring. Eventually, these thermoscopes become trusted to the point that they produce new knowledge about the world, as with Sagredo's observations. After this, numerical markers are added to what are now basic thermometers. These markers are arbitrary and singular at first, but then tied to fixed points, allowing temperature scales to be replicated and data to be shared. As technical advances improve the performance of these thermometers, the fixed points can be narrowed to greater degrees of precision, variations identified and accounted for. The important thing is that each successive stage builds upon previous knowledge, each new level supported by those below. And slowly, over centuries of work by hundreds of individuals, a reliable scale is built, degree by degree, out of thin air.

Reaching absolute zero

This story does not end with the creation of reliable thermometers, though, and temperature contains more nuances than can be located on Celsius's and Fahrenheit's scales. This was readily apparent to William Thomson, or Lord Kelvin, the nineteenth-century British physicist who claimed that 'when you can measure what you are speaking about, and express it in numbers, you know something about it'. It was Thomson's belief in metrology, and his application of these methods to temperature, that would push our understanding of hot and cold into modes that recapture the grandeur and scale of the ancient Greeks' philosophy.

Thomson's problem with the measurement of temperature was that it was always second-hand. No matter how reliable thermometers were or how sensitive their readings, they relied on an intermediary: the fluid inside the instruments that expanded and shrank in response to changes in temperature. These scales didn't measure temperature directly, nor was there any method to definitively relate readings from different scales to one another. As Thomson complained in a paper in 1848: 'Although we have thus a strict principle for constructing a definite system for the estimation of temperature', the scales themselves are simply 'an arbitrary series of numbered points'.[31] They didn't address the heart of the matter – the actual nature of temperature – any better than the thermoscopes of Galileo and Santorio.

Part of the reason for this uncertainty was that people in Thomson's day had no clear explanation of what heat actually was. For centuries, alchemists, natural philosophers, and scientists had debated whether or not heat was a substance in its own right, a distinct element that existed within matter, or whether it was a behaviour exhibited *by* matter, most commonly thought of as a form of motion. Both approaches explained some aspects of temperature but neither was definitive, and the importance of temperature in other fields, like chemistry, meant resolving the issue was a pressing concern. As the Dutch physician and chemist Herman Boerhaave commented in 1720: 'If you make a mistake in your exposition of the Nature of Fire, your error will spread to all the branches of physics, and this is because, in all natural production, Fire [...] is always the chief agent.'[32]

In the eighteenth century, the concept of heat's physical existence was conveyed by what is called 'caloric' theory, referring to an invisible substance of the same name that was thought to flow seamlessly throughout matter. The eighteenth-century French chemist Antoine Lavoisier, who often drew on this theory in his experiments, describes caloric as 'exquisitely elastic' and 'very subtle', a weightless fluid that can penetrate any material and is strongly attracted to normal matter, capable of 'insinuating itself between the particles of bodies' as quickly as a shadow.[33] Like many of his contemporaries, Lavoisier was not fully committed to caloric's existence, but the theory helped guide experiments and account for certain phenomena. Caloric, which was thought to be self-repelling, was used to explain why heat flows from hot to cold objects, for example, and why metal expands when warmed – it was being stretched from the inside out as it filled with caloric.

The development of reliable thermometry brought new evidence to this debate. Lavoisier himself had already helped banish an earlier theory that posited the existence of a substance called phlogiston, from the ancient Greek for 'burning up', that was thought to be released into the air when objects were burned. Unlike caloric, phlogiston had mass, but Lavoisier disproved this theory, in part by showing how some substances gain weight when burned. (This would eventually lead to the discovery of oxygen as the key element in combustion.) With caloric, one problem that arose from the quantification of temperature was that of latent heat: energy that is released or absorbed when matter changes from one state into another, like gas into liquid. With sensitive thermometers, scientists could observe that warming a bucket of melting ice did not change its overall temperature until the ice had completely melted, even though caloric theory predicted that this change should occur as soon as the ice began melting and caloric started seeping into the liquid. Proponents of the theory found a way to adapt it to such challenges, but it was one of a growing number of inconsistencies. As experimental evidence accrued, confidence in caloric theory waned. The problem was that a definite alternative had yet to emerge.

––––––

The new theory of heat that would be outlined by Thomson and many of his peers was ultimately inspired by one of the icons of the age: the steam engine. It was these machines that would help provide proof of something that scientists dating back to Francis Bacon and Galileo had long suspected: that heat is a form of motion. It was this theory – which Thomson would christen thermodynamics, first using the term to refer to steam engines themselves as 'thermodynamic engines'[34] – that removed caloric from our understanding of the world.

Thomson was inspired in his work by a text named *Réflexions sur la puissance motrice du feu* ('Reflections on the Motive Power of Fire'), which was published many decades earlier in 1824 by French scientist and military engineer Nicolas Léonard Sadi Carnot. It was the only book Carnot ever published, and was ignored almost completely by his peers, but it contained enough original thinking for a lifetime's work. 'It was utterly without precedent and dense with implications,' as one historian puts it.[35]

Carnot was primarily interested in improving the efficiency of steam engines, seeing them as a tool of military and economic might. His father had been war minister under Napoleon, and Carnot understood that it was the UK's steam engines, not its armies or navies, that ensured its victories in Europe. Like Aristotle's understanding of the human body, in which the heart distributes the heat that animates the limbs, Carnot saw the steam engine as the motive force in the country's economy: a machine that pumped coal and iron through industrial arteries, producing ships, and rifles, and ammunition in the process.[36] In his *Réflexions*, Carnot tried to better understand the steam engine by abstracting its working. He removed the distractions of riveted iron and leaky steam and focused on the underlying properties of the machine. He noted that these engines worked by transferring heat from a hot body to a cold one, then transforming this into movement. Carnot compared the mechanism to that of a watermill, which captures the flow of water running downstream to power a turbine. Although Carnot himself believed in the caloric theory of heat, by simplifying the operation of steam engines in this way he described what would eventually become the second law of thermodynamics: the inescapable flow of heat to colder systems.

In the age of steam, full of pumping engines fed by fiery furnaces, Carnot was not the only person to spot a connection between temperature and motion. Across Europe, scientists puzzled over different aspects of this problem, including the German physician Robert Mayer and English physicist James Joule – two individuals who worked independently on what would become the first law of thermodynamics. This is the conservation of energy, or the notion that energy cannot be destroyed, only transformed.

It was Joule who gained most credit for his work, thanks in part to his ability to encapsulate his ideas through experiment. He was primarily concerned with the notion that heat and motion are not just connected but interconvertible: capable of being changed back and forth (something caloric theory did not allow). Joule set out to prove this idea like a painter obsessed with a particular scene, illuminating it over and over from different angles. His best-known demonstration was an apparatus that consisted of a paddle attached to a weight on a pulley and suspended in a flask of water. As the weight fell, the paddle rotated and the water was heated. Not only did the device illustrate that heat could be created from motion (in this case, from the friction of the paddle in the water), but more importantly it allowed Joule to quantify the process – to measure it. By recording temperature changes in the water, the mass of the weight and the distance it travelled, he could calculate heat as a form of motion, or work. To be precise, he noted that, using his apparatus, it took around 780 foot-pounds of effort to heat one pound of water by one degree Fahrenheit.[37]

Thomson began to be convinced by the idea that heat and motion were equivalent after listening to a presentation by Joule on the matter. He later said he 'felt strongly impelled at first to rise and say that it must be wrong', but as he sat and listened to Joule, he realised he'd discovered 'a great truth [and] a most important measurement to bring forward'.[38] What continued to bother Thomson was the idea that you could *completely* transform heat into motion, or work. Plainly, there were situations where this was not the case. 'When "thermal agency" is thus spent in conducting heat through a solid, what becomes of the mechanical effect which it might produce?' he wrote in 1849. 'Nothing can be lost in the operations of nature, no energy can be destroyed. What effect then is produced in place of the mechanical effect which is lost?'[39]

It was problems like these that led Thomson and German physicist Rudolf Clausius to formulate what has now become the familiar explanation of the mechanics of heat. They drew upon a concept that seems synonymous with science today, yet had only been tentatively introduced at the beginning of the nineteenth century: energy. We now understand energy to be the 'capacity for work', a broad and flexible concept that describes a variety of phenomena in the natural world. But for those looking to explain the new science of thermodynamics, it would come to mean to something much more specific: the motion of atoms. It was this energy, in the form of molecular motion, that ultimately constituted heat. It explained why heat could not be transformed into work without loss: heat is a 'degraded' form of energy, much of which is unorganised and random. This is what leaks off into the world, and which Clausius described as 'entropy' – a measure of disorder in the universe. In 1865, Clausius captured this new science with brutal economy, as the two laws of thermodynamics:

1) The energy of the universe is constant
2) The entropy of the universe tends to a maximum[40]

These maxims would have a profound effect on a range of scientific disciplines, but the formalisation of this new theory of heat, of thermodynamics, also allowed Thomson to finally overcome the shortcomings he saw in the measurement of temperature. Not only could he now measure heat using abstract calculations of the amount of work produced, he could also define a new base for the temperature scale. If heat was a factor of atomic motion, then surely that meant there could be such a thing as total absence of motion, of energy, and of heat. Thomson described this as 'absolute zero' – the point at which, well, nothing happens. He formulated his new measure of temperature as degrees Kelvin, which has since become the standard unit for much scientific and high-precision work. On this scale, absolute zero is defined as 0K, or $-273.15°C$, or $-459.67°F$. Although it's physically impossible to reach absolute zero, Thomson had finally established a measure of temperature that was not dependent on the behaviour of some specific substance. After centuries of experimentation, thermometry had finally anchored itself in something indisputably solid.

Everlasting fire; energy and entropy

With the quantification of temperature in the eighteenth century, scientists were 'quite prepared to sacrifice rich concepts in order to promote rigor and clarity', writes historian of measures Theodore Porter. Their 'infatuation with measuring led to the neutralisation of concepts as well as their creation.'[41] The metrologists had quenched Heraclitus's cosmic fires and replaced them with a notion of heat filtered through thermometers and dripped into thin scales. This claim, that measurement diminishes its subject, is one that appears frequently in discussions of metrology. It is part of a broader charge laid against the sciences and captured best in Max Weber's concept of disenchantment: the displacement of the supernatural by the scientific and an accompanying loss of meaning. Think, for example, of how caloric – a substance of almost folkloric properties: weightless, frictionless, and invisible – is supplanted by the brute mechanics of the steam engine and the dynamic theory of heat. In this framework of knowledge, the mysteries of the universe are eliminated as chemistry replaces alchemy and enchantment is subdued by engineering. Or so the story goes.

I don't agree that our changing understanding of temperature has led to an impoverishment of meaning. Instead, just as the heat of the furnace turns coal into steam and the motion of turbines, the scientific refinements of thermometry have transformed the richness of their subject, revitalising old mythologies and offering new ways to understand the world.

Consider, for example, how the instrumentation of thermometry has furnished our language with new explanatory ideas. In the eighteenth and nineteenth centuries, the thermometer and its cousin the barometer were eagerly adopted as metaphors, appearing in books, newspaper articles, and political speeches. The visual language of falling and rising liquid was intuitively understood, while the uncanny sensitivity of these devices to invisible phenomena suggested they might have other, mysterious powers of quantification. Lavoisier spoke of 'the thermometer of public prosperity',[42] while satirical printmaker William Hogarth used thermometers in his illustrations to indicate that passions, amorous and religious, were running high. A barometer manual from 1694 tells readers explicitly that they are capable of measuring not just pressure but human temperament: 'the lower the *Quicksilver* descends, the more listless and out of order Men's Bodies are, because the Air is then full of that which is disagreeable to the Nature of Man, who was not made to live in a Watry Element'.[43]

Thermometers' use of quicksilver, a silvery, slippery chemical long associated with transformation and mobility, established this connection to spirit, writes American literary critic Terry Castle. In her essay 'The Female Thermometer', she tracks how these instruments are associated with erotic passions and the female temperament, becoming new expressions for the era's misogyny. One satirical essay published in 1754, for example, describes the invention of a thermometer that can measure 'the exact temperature of a lady's passions'. Its degrees run from 'Abandoned IMPUDENCE' at the top to 'Inviolable MODESTY' at the bottom, and its owner uses it to probe the morality of various London locales. In a gruesome piece published in *The Spectator* in 1712, the quicksilver in a thermometer is replaced with 'a thin reddish Liquor' extracted from the heart of a dead 'coquette', which turns the device into a dowsing rod for her departed passions. 'The instrument proves to be as man-crazy as the woman who provides its vital fluid,' writes Castle. 'It danced up,' recalls its inventor, at the approach of "a Plume of Feathers, an embroidered Coat, or a Pair of fringed Gloves" and dropped abruptly in the presence of an "ill-shaped Perriwig, a clumsy pair of Shoes, or an unfashionable coat."[44]

The concepts of temperature uncovered by these quantitative methods also have their own impact. Awareness of thermodynamics in particular had a huge effect on nineteenth-century thought, shaping social and political arguments just as Darwin's theory of evolution had. It's no surprise when we consider how the core elements of the theory – the two laws of thermodynamics in particular – have a sense of mythology that connects with ancient intuitions about the world. The first law, the conservation of energy, offers constancy and endurance: the reassurance that nothing is lost forever. Its promise is then undercut by the second law, the increase of entropy, which warns that what is preserved by the universe will not be life as we know it, only disorder. This sense of balance was intended by Clausius himself, who wrote that he 'intentionally formed the word *entropy* so as to be as similar as possible to the word *energy*'. His purpose was to reflect the symmetry of these 'two magnitudes' – these sublime concepts.[45] It is energy that provides the dynamism of the universe, continually transforming the world around us, and it's entropy that ensures that all we see will eventually dwindle. For many in the nineteenth century who watched the landscape around them transformed through the energies of the steam engine, railway, and factory, these truths may have seemed self-evident. Of course there is progress and order in the world, they would say; of course there is wastage and decay.

To completely map the effects of thermodynamics on culture, religion, and philosophy is a desperate task, but one particular aspect can perhaps serve as a sample of the whole: the theory of the heat death of the universe, a natural consequence of the second law of thermodynamics. As Thomson explained in a popular magazine article in 1862, 'although mechanical energy is indestructible, there is a universal tendency to its dissipation, which produces gradual augmentation and diffusion of heat, cessation of motion, and exhaustion of potential energy through the material universe'. The consequence is that even a huge, apparently inexhaustible battery like the sun must eventually run down, and 'inhabitants of the earth can not continue to enjoy the light and heat essential to their life for many million years longer unless sources now unknown to us are prepared in the great storehouse of creation'.[46]

For Thomson himself, this idea was not unappealing. He was a committed Christian and felt that a finite and unidirectional cosmos was in keeping with biblical scripture.[47] Just as man's life had a beginning and an end, with a fated hereafter, so too should the entire universe eventually shut up shop and tally the accounts. More than once in his writing he compares the laws of thermodynamics to Psalm 102, which seems like a presentiment of the same balance of constancy and wastage that Clausius described: 'They shall perish, but thou shalt endure: yea, all of them shall wax old like a garment; as a vesture shalt thou change them, and they shall be changed.'[48] What were thermodynamics but a restatement of scripture?

Not everyone was happy with this interpretation, though, and scholars have suggested that the apparent confirmation of a cosmic end point by scientific diktat 'triggered a widespread cultural anxiety'[49] (another interpretation: thermodynamics simply provided a new vehicle for existing social anxieties[50]). Some attempted to neutralise the threat with a new synthesis of science and religion. The 1875 bestseller *The Unseen Universe*, written by a pair of physics professors, 'reimagine[d] energy physics as a site of transformative spiritual consolation', presenting the continuity of the first law of thermodynamics as only a new sort of afterlife.[51] Others reckoned with their fears through art, with the dying sun proliferating as a literary metaphor. Algernon Swinburne's 1866 poem 'Garden of Proserpine' concludes with this image of total stillness: 'Then star nor sun shall waken, / Nor any change of light [...] Only the sleep eternal / In an eternal night.'[52] Other times, the metaphor is tied to other forms of dissolution and dissipation, as in Joseph Conrad's 1899 novel *Heart of Darkness*, where repeated descriptions of sinking red suns accompany the narrator's nightmarish trip into the colonial brutality of the Congo Free State. Conrad was more explicit about such pessimism in his personal letters, writing in one: 'If you believe in improvement you must weep, for the attained perfection must end in cold, darkness, and silence.'[53] Such laments hardly rely on thermodynamics for their inspiration (images of the sun being extinguished can be found in plenty of ancient texts) but were given new authority and relevance with these scientific discoveries.

Fears of this ultimate quiet and darkness seem to have been particularly acute in an age when technology had delivered such a glut of artificial sound and light. And fittingly, it was within the science-fiction genre that entropy and the heat death of the universe were most obviously explored. The most famous example is H. G. Wells's 1895 novella *The Time Machine*, in which our protagonist, known only as the Time Traveller, journeys forward through Earth's future to witness the end of everything. He stops first in the year 802701 and discovers that a new class system has taken hold on the planet, with the beautiful and childlike Eloi gambolling on the planet's surface, while the murderous and inhuman Morlocks plot underground. After a narrow escape from the Morlocks, the Traveller skips forward further to witness the mighty clock of the cosmos winding down around him. 'At last a steady twilight brooded over the earth,' notes the Traveller. 'The circling of the stars, growing slower and slower, had given place to creeping points of light. At last, some time before I stopped, the sun, red and very large, halted motionless upon the horizon, a vast dome glowing with a dull heat, and now and then suffering a momentary extinction.' He slows his machine and finds himself stranded on a 'desolate beach' at the end of time, inhabited only by huge and decrepit creatures – giant crabs with blemished and cor-rugated shells and huge white butterflies, flapping lazily overhead. Appalled but unable to stop himself, the Traveller pushes further forward, millions of years into the future, until, finally, even the sun is extinguished and nothing else remains:

> All the sounds of man, the bleating of sheep, the cries of birds, the hum of insects, the stir that makes the background of our lives – all that was over [. . .] The breeze rose to a moaning wind. I saw the black central shadow of the eclipse sweeping towards me. In another moment the pale stars alone were visible. All else was rayless obscurity. The sky was absolutely black.[54]

The stillness that Wells describes lingers long after his story has fin-ished, and the vision he offers has become one of the most resonant apocalypses in the secular world. Quantification, measurement, and the scientific method may have banished certain myths from nature, but they've also introduced their own strange visions, as inaccessible as any magical realm yet just as real to our imaginations. Heraclitus's everlasting fire no longer roils eternally through the cosmos, but instead, atoms and molecules jostle invisibly throughout the world with their own energy and heat. Until, at last, they stop.

5

THE METRIC REVOLUTION

---

*The radical politics of the metric system and
its origin in the French Revolution*

Then will arrive the moment in which the sun will observe in its course free nations only, acknowledging no other master than their reason.

—MARQUIS DE CONDORCET[1]

The Metre and Kilogram of the Archives

You might expect the corridors of the Archives nationales in Paris to smell a little musty. After all, here in the Hôtel de Soubise, a grand city mansion requisitioned by the state during the French Revolution, the receipts of a nation are piled high. Rooms that once hosted aristocratic balls are now filled with stamped and numbered cardboard boxes, packed with records of state and tucked away neatly into ornate multi-storey bookshelves serviced by iron walkways and stepladders. The atmosphere is quiet and expectant, as if the building were simply waiting for the present to slip into the past and find a place in its catalogue. The air, though, surprises: it's neither stale nor damp but delicately fragrant. On this spring morning, it even carries the faintest trace of honey.

'It's the wax,' explains Sabine Meuleau, a curator at the archive who is currently leading us through doors marked '*Accès interdit*'. 'They use it on the floors and on the shelves.' She pauses for a second and sniffs the air, trying to catch the ghost of a scent she's long become accustomed to.

As we stalk through the building, we draw nearer to our destination: the armoured heart of the archive – a huge iron cabinet that contains its most precious documents. Behind three sets of locked doors and an alphabetic combination set in 1791 (and never changed since), there are treasures aplenty. Here is the pledge signed by the Third Estate in 1789, the famous Tennis Court Oath, where the commoners swore not to be moved 'until the constitution of the realm is established'. Here is the original copper engraving of the *Declaration of the Rights of Man and of the Citizen*, adopted that same year, with its epochal assurance that 'men are born and remain free and equal in rights'. And here is the last letter of King Louis XVI, who was guillotined on 21 January 1793 under the desacralised name of Citizen Louis Capet, an act that drew a sharp and irreversible distinction between the Ancien Régime and the new republican age. In short, here is the history of perhaps the greatest revolution of the modern world.

We're not looking for paper, though. After Sabine and her colleague Stéphanie Marque-Maillet have placated the various locks and wards of the cabinet, they retrieve two packages from its interior: one long and thin, the other small and octagonal. Inside are objects that speak as eloquently about the ideals, failures, and successes of the Revolution as any document. These are the original standards of the metric system, a method of measurement inaugurated alongside the new republican order and just as revolutionary. These are the Metre and the Kilogram of the Archives.

They are unassuming things to look at, nestled within faded red velvet cases on the table in front of us. Slim, shiny, and entirely unmarked, they remind me of the first Celsius thermometer: a silvery icon that is intimately familiar in shape. Unlike the measuring standards of ancient kings and queens, they carry no seals or crests to confer their authority but derive their power from the rationality of their design. Instead of the monarch's orb and sceptre, this is the weight and length of the revolutionary. As I gaze at the pair of standards, drawing closer and closer to their gleaming surfaces, I'm told by Sabine – politely, firmly, repeatedly – to look but not touch.

The history of the metric system, the world's most successful system of measurement, is bound up in the politics of its creation. It was designed by the French intellectual elite of the eighteenth century, the *savants*, to replace the arbitrary units of measurement used during the Ancien Régime. Instead of the *pied du Roi*, or 'king's foot' (a linear measure dating back to Charlemagne), units would be defined by an impartial and unchanging arbiter, 'the greatest and most invariable of bodies which man can measure':[2] the Earth itself. The metre's length would be a fraction of the planet's meridian, an imaginary line running from the North Pole to the South, while the kilogram would be defined as the weight of 1,000 cubic centimetres of water. Such definitions would transcend the egos of kings and idiosyncrasies of nations. These would be weights and measures *à tous les temps, à tous les peuples*: for all times and for all people.

Even the raw materials of the metre and kilogram embody this history. Both are cast from platinum, a metal that had only recently been discovered by Europeans during colonial expeditions to South America, prized for its purity and toughness. Contemporary reports note that it was unmarked 'when struck on an anvil of steel' and could be extracted 'only with infinite labor and charge'.[3] Using this material demonstrated human mastery over nature. And even if these originals were destroyed by war or wear, clever minds would be able to reconstruct them to their exact same values using their scientific definitions. As Napoleon Bonaparte would later declare of the metre: 'Conquests will come and go but this work will endure.'[4]

None of which is quite true, though. France would reject the metric system not long after its introduction (only to adopt it once more decades later), while modern satellite surveys have shown that the *savants'* measurement of the meridian is not quite correct. Errors made in their calculations mean the metre of the archives is actually 0.2 millimetres (or 0.008 inches) short of its target value.[5] It's a difference in thickness of just a few sheets of paper, but one that has been perpetuated in every metre ever since.

The artefacts in front of us in the Hôtel de Soubise are the product of high-minded and scholarly work, but demand for standard measures in France started not with the *savants* but with the commoners. The need to reform the country's weights and measures was a major complaint in the run-up to the French Revolution, a single irritation among the many economic and bureaucratic failures that set the stage for political upheaval. Journeying through France in 1789, the English agriculturist and travel writer Arthur Young commented on the 'tormenting variations' of the country's units of measurement. '[T]he infinite perplexity of measures exceeds all comprehension,' he wrote. 'They differ not only in every province, but in every district, and almost in every town . . . The denominations of the French measures, as the readers will see, are almost infinite.'[6]

It wasn't just France, of course; units of measurement in Europe at this time were both elastic and profligate. There were measures fitted for every profession and purpose, units that would shrink and swell to better match their target. Such elasticity could be functional, providing a little give in the bonds of the feudal system; but in Ancien Régime France, where serious metrological reform hadn't been undertaken in centuries, the situation was particularly dire. The capacity unit known as the *pinte*, for example, measured 0.93 litres in Paris, 1.99 litres in Seine-en-Montage, and 3.33 litres in Précy-sous-Thil. The most common measure of cloth, the *aune*, could refer to one of seventeen national variations, ranging in size from roughly 300 to 600 *lignes*.[7] And how long was a *ligne*? Well, about 2 millimetres, but again, it depended. One estimate suggests there were more than 1,000 recognisable units in France at the time, and a stupefying 250,000 local variants.[8]

All this meant that in 1789, when the Estates General of France were called during the prelude to the Revolution, a demand to standardise weights and measures was high on the agenda. In the huge survey of national grievances intended to guide reform, known as the *Cahiers de doléances*, the issue is mentioned by members of the Third Estate more frequently than complaints about courts or infringement of personal liberties.[9] 'The consistent unity of the nation's attitude towards the problem of standardisation of weights and measures is astonishing,' says Witold Kula. 'Everybody wanted it, from peasants in different regions to the craftsmen in diverse trades to municipal leaders of virtually all towns.'[10] Commoners wanted to protect themselves against the chicanery of their masters, while the masters wanted it to spur trade and agricultural output. The desires of these two groups were not always aligned, especially when the grandiose plans of the *savants* actually made contact with the needs of everyday workers, but in the *Cahiers* at least they were united by a single demand that's repeated time and time again: 'One king, one law, one weight, and one measure.'[11]

The metre and the kilogram of the archives are the product of these very different demands; of the idealism of the *savants* and pragmatism of the peasants. As artefacts, they capture this sense of duality: they are intensely engineered objects that aspire to a state of nature; products of rational calculation as arbitrary as the units they replaced; and supposedly egalitarian tools that were designed for and benefited the elites. But despite all these contradictions, they worked.

As we stand in a beam of sunshine that's broken through the clouds outside to light the surface of the metre and the kilogram, Stéphanie notes that for all their beauty and historic importance, these artefacts don't really draw a crowd when they go on display. On the rare occasions they're taken out of their iron home, they're often overlooked by visitors, unaware of their significance. 'People have seen metres all their lives and they wonder why we're showing them,' she says. 'They don't understand why it has to be an object because it's already in their heads.'

The irregular Earth

On 9 March 1790, the politician, aristocrat, and sometime bishop Charles Maurice de Talleyrand-Périgord stood before the National Assembly – the newly christened political body that now declared it represented the people of France – and told them they lacked ambition.

In the last year alone, revolutionaries had stormed the Bastille, abolished the feudal rights of the nobility and clergy, and affirmed the Declaration of the Rights of Man. But while the French people were now equal before the law, said Talleyrand, they still suffered under the inequalities of weights and measures. Every day, honest men and women were beaten down by the iron *toise* of the Ancien Régime and dealt short shrift with every *livre* of grain poured into their bowls. 'The great variety to our weights and measures occasions a confusion in our ideas, and necessarily an obstruction to commerce,' said Talleyrand. 'Nothing can justify such an abuse,' he concluded, and only 'a reform in weights and measures' could relieve it.[12]

What Talleyrand said was common sense, as the *Cahiers de doléances* had proved. But while most proposals to fix the country's weights and measures argued for the nationwide adoption of Parisian standards, Talleyrand's went further. Adopting Parisian measures failed to rise to the 'importance of the matter, nor the aspirations of enlightened and exacting men.'[13] Guided by the *savants* of the Académie des sciences, he laid out an ambitious plan for the reform of France's weights and measures. Consistency was a start, but the Assembly should aspire to greater things and create an entirely new system of measurement derived from the precepts of nature. Doing so would follow the highest ideals of the Enlightenment, while distinguishing France as a global leader in science. 'This plan, simple and perfectly exact, must unite all opinions and even excite amongst the learned nations, a laudable rivalry,' said Talleyrand.[14]

The metric system had found its perfect salesperson. Talleyrand was a creature of uncanny political instinct, able to read the turbulent flows of the Revolution like a ferryman, finding safe passage where others drowned (even if that meant jettisoning companions he'd championed moments earlier). A contemporary historian notes that, during his life, he betrayed kings and emperors but never France.[15] Napoleon just called him a 'shit in a silk stocking.'[16] In 1790, though, Talleyrand had the foresight to see that the political climate demanded radical, not incremental, change. Anything less was a betrayal of the people and of the opportunity they'd created through their blood and suffering. As he told the Assembly: 'When a nation is resolved to establish a great reform, it is necessary that it should avoid, nay, even that it should be cautious not to do the work by halves, lest it should be obliged to perform it over again.'

The Académie's recently formed Commission of Weights and Measures shared this ambition, seeing the Revolution as a chance to remake France through measurement. Its changing roster included some of the world's greatest scientists, from Joseph-Louis Lagrange to Pierre-Simon Laplace and the aforementioned Lavoisier. Foremost among them for metrological ambition, though, was the Marquis de Condorcet, a philosopher and mathematician who guided Talleyrand in his presentation to the government, and whose personal ideology would be captured in the platinum standards of the archive.

In the months and years following Talleyrand's speech, the Commission fleshed out their plan for the metric system, settling on a handful of important characteristics. Firstly, they decided that the units should be interconnected. The capacity measure should be constructed from a cube made of the length unit, which, when filled with water, would provide the basic unit of weight. These bonds would mutually enforce each unit's value – no individual measure could fall into imprecision when held in place by its brethren – while simultaneously making calculations and conversions between units easier.

Secondly, the system should be decimal, with all units divisible by 10. This was, and is still today, a controversial demand, with opponents arguing that base-12 and base-16 systems (like those used by British imperial and US customary measures) are easier for calculation. They allowed users to divide units into halves, thirds, and quarters without resorting to decimal places, simplifying daily transactions. But the *savants* had decided that decimalisation was ready for a larger audience. It had long proven its worth in mathematics, and advocates like Lavoisier argued that if introduced to measurement, it would bring similar benefits to fields like engineering and commerce. It was a belief that reflected the ambitions of the Enlightenment. Science was reaching out across the cosmos and delving into microscopic realms, and while halves, thirds, and quarters were useful for haggling in the market, only decimalisation, with its ability to enumerate both the very large and very small, could satisfy the era's ambition.

The third major change was linguistic. The *savants* would devise an entirely new taxonomy for the system, with new names for units paired with Greek and Latin prefixes to denote multiples and fractions. These included familiar terms like *kilo* for 1,000 units and *cent* for 0.01 parts, but also less common terms like *demi* for a half unit and now forgotten prefixes like *myria*, for multiples of 10,000. In 1790, the neologism 'metre' was coined for the basic unit of length, derived from the Greek *metron*, meaning both something used to measure and measures in poetry and music. It was a name 'so expressive that I would almost say it was French', enthused its creator, mathematician Auguste-Savinien Leblond.[17]

One problem remained. While script and scenery of the metric system was now prepared, its star turn had yet to arrive. The new length unit, the metre, would be the keystone of the metric system, and while several proposals had been made for its definition, the *savants* were still wavering between two. They would either derive its length from a fraction of the Earth's meridian or use a seconds pendulum: that is, a pendulum with a back and forth swing that lasts exactly two seconds. Despite the confidence of Talleyrand and the Commission in the wisdom of units derived from nature, neither definition was going to be an easy sell.

For many years, it seemed like the seconds pendulum would be the preferred method. It had backers in England and France, and traced its lineage back to the work of Galileo, who, in 1602, showed that the time it takes for any pendulum to complete a full movement back and forth (known as its period) never changes, regardless of how wide it swings.[18] This is because pendulums swing faster during longer arcs and slower during narrow ones, a result of the different exertions of gravity on the weight. The benefit for the *savants* was that this means the length of a pendulum is linked to its periodicity. If you can measure one of these factors, then the other will always be the same. Like the boiling and freezing points of water, this property, known as isochronism, is one of those convenient oases of stability we've discovered in an otherwise chaotic universe. A seconds pendulum is one with a period of two seconds, with a second itself defined as 1/86,400th of a solar day. This means it has a length of 39.1 inches or 993 millimetres – less than a centimetre short of today's metre. All you would need to replicate this unit was a line, a weight, and a clock.

Despite the elegance of this solution, the seconds pendulum had its flaws. Firstly, the swing of a pendulum is not, strictly speaking, as consistent as Galileo had suggested, varying with latitude and the observer's distance from the equator. In 1672, a member of the Académie des sciences had discovered the swing of his seconds pendulum in South America was 1.25 *lignes* (2.8 millimetres) shorter than in Paris.[19] At the time, the finding baffled the scientist, but was seized upon happily by Isaac Newton, who used it as evidence in his 1687 work *Principia* to show that the Earth could not be a perfect sphere due to the effects of gravity, a then controversial theory.[20] Because of centripetal forces, predicted Newton, the Earth would be flattened at its poles, contributing to the decrease in gravity near the equator in South America that slowed the pendulum's swing.

For the *savants* of France, this meant any unit of measurement derived from the seconds pendulum would only be true at a certain latitude, a stipulation that would fasten the definition to a specific place, favouring some nations more than others. And even then, gravity within a single latitude could vary due to the presence of particularly large objects like mountains. An additional worry was that the pendulum definition would link the new unit of length to the value of the second, and at the time, the second's continuing existence was not a given. It, too, was being considered as a candidate for reform – part of a plan to redefine the 60-second minute and 60-minute hour that Europe inherited from ancient Babylonian astronomers. The *savants* did not want to tie their hands unnecessarily.

This left the meridian as the best source for the new metre, though it too had its problems. In the century since Newton had published his *Principia*, the scientific community had agreed, after much discussion and several costly geodesic expeditions, that the Earth was indeed an oblate spheroid, flattened on both ends. But exactly how flattened it was, and how much its curvature varied, remained contentious. If the metre was to be defined as a fraction of the meridian – an imaginary line running through the North and South Poles – then the meridian itself would need to be measured anew. Given the changing nature of the Earth, which had been reckoned a perfect sphere just a century before, surveys from earlier years would not suffice. In the end, the Académie selected two of their number to do the job: the astronomers Jean-Baptiste-Joseph Delambre and Pierre-François-André Méchain. The pair wouldn't survey the full meridian, but a section of its arc running straight through Paris: from Dunkerque, on the northern coast of France, to Barcelona, just over the Spanish border. Like tailors measuring a gigantic client for a suit, the pair would crawl over this expanse of territory, one headed north and one south, before multiplying the distance to calculate the half-meridian. They would then divide this length by 10,000,000 to create the definition of the metre.

The expedition itself took seven years, during which Delambre and Méchain struggled not only with the demands of precision, but also with the fevered climate of the French Revolution. Their method of survey was triangulation, which uses geometry to calculate distance. Their approach was based on the Euclidean principle that if you know the three angles of a triangle and the length of one side, you can calculate the length of the other two. So, after measuring out two baselines at either end of the country to create their first known lengths, the astronomers then connected these using a series of readings taken from high vantage points. Imagine it as a giant game of connect the dots, with the pair climbing church steeples, fortresses, and hills to measure the angle from one mark to the next.

By connecting all these points, they could calculate the full distance between their baselines.

As relayed in Ken Alder's classic of metrological history, *The Measure of All Things*, this unfamiliar activity meant they were often challenged by locals, who took them for spies, or worse, counter-revolutionaries. 'At every turn they encountered suspicion and obstruction,' says Alder.[21] It did not help that the commission authorising their trip was issued in the name of King Louis XVI, who had been executed in 1793, a year into their journey, or that they looked and acted like land surveyors – individuals whose appearance, measuring fields and property, often preceded an increase in taxes. In one encounter in Saint-Denis, Delambre was detained by a mob of town folk suspicious of his motives. Earlier in the Revolution, the assistant to the town's mayor had been stabbed fourteen times for refusing to lower the price of bread.[22] Delambre wrote later that he was forced to give an impromptu lesson on geodesy to the suspicious crowd, who had likely demanded in the *Cahiers de doléances* the very reforms he was attempting to execute. Delambre recalled that his audience was 'quite large,' and that 'impatient murmurs began to be heard; a few voices proposed one of those expeditious methods, so in use in those days, which cut through all difficulties and put an end to all doubts.'[23]

The outcome of these years of danger and labour was a measurement of stunning precision that, as Alder puts it, also 'invalidated the guiding premise of the entire mission.'[24] The *savants* of the Académie had chosen to define the metre using the meridian because they believed it was perfect and unchanging, but the astronomers' patient trawl revealed that the opposite was true. Their measurement showed that not only was the Earth *not* a perfect sphere, it was not even a regular ovoid. Their net of triangles, stretched over the hills and mountains of France, revealed the planet's surface to be rucked and contorted – as uneven as shrivelled fruit. It is the sort of revelation that appears so often in the history of measurement, as the quest for precision – the desire to burrow ever more closely into the weft of reality – unveils only irregularity on a far greater scale.

Far from being dismayed by the news, Méchain was stimulated by the discovery. He marvelled at the intransigence of the Earth, which 'refused to conform to the formulas of my mathematical colleague', and in private letters expressed a sense of disbelief tinged with awe. 'Why did He who molded our globe with his hands not take more care . . .?' he asks in one. 'How did it happen that by the laws of motion, weight, and attraction, which the Creator presumably decreed before He set to work, He allowed this ill-formed earth to take this irregular shape for which there is no remedy, unless He were to begin anew?'[25] Like a child who discovers for the first time that their parents are not infallible, he seems to have found a dangerous thrill in the error.

When in 1798 the data from the survey was finally compiled, the eccentricity of the Earth was softened in order to create a measurement that the *savants* thought best captured their understanding of the planet's overall curvature.[26] They had already alienated colleagues in Britain and America by using the Paris meridian for the metre's definition, and did not want to discourage any further those nations who otherwise approved of the metric system. The final data also ignored a small discrepancy in the measurements taken by Méchain, perhaps created by a fault in his instruments. Méchain concealed this error during his lifetime, but its existence tortured his conscience. It was only revealed in a quiet and respectful fashion after his death, when Delambre wrote the official history of the expedition, a 'calculated homage to the transience of human knowledge', according to Alder.[27] That means the metre bar contains not just one error (the slightly short meridian length, as proved by modern surveys) but two. Neither matter in the slightest.

The unveiling ceremony for the new prototypes took place on 22 June 1799, when the final platinum metre bar – now *the* definition of the metre – was presented to the Council of Ancients. It was housed in the legislative chamber, 'just as the Athenians had kept their measures in the Acropolis and the Israelites kept theirs in the Temple',[28] to underscore the metre's new authority. It was and is a modern relic: a superlative object with a tragic flaw that was forged to embody a higher order but nevertheless encoded human error. As Laplace said during the ceremony, the creation of the metre meant that every humble farmer in France could now say: 'The field that nourishes my children is a known portion of the globe; and so, in proportion, am I a co-owner of the World.'[29] The irregular but awe-inspiring world.

Ideology and abstraction

It was not his intent, but Laplace's speech in 1799 unwittingly highlights tensions at the heart of the metric project – tensions that would contribute to its downfall and that still plague the business of measurement today. Can the metre really be said to be natural just because it is a 'known portion of the globe'? What's natural about a number determined by a small group of experts at great effort and expense? And why should the people of France even care about these origins? Did they feel like 'co-owners of the globe', or was Laplace's grandiloquence aimed at his peers? After all, it was the scholars and scientists who had girdled the planet, shrinking its span into a platinum length that could be held in the hands. No king could boast of greater conquest.

Like Enlightenment thinkers in general, the metric *savants* were varied in their politics: some incrementalist and some militant; some conservative and others radical. But all shared a few key ideals, foremost among which was a belief in the ability of humanity to shape its destiny through rational means. For the *savants* and many other eighteenth-century elites, the scientific discoveries of past decades proved that empirical methods could not only reveal new truths about the universe, but also reorder the cosmos itself, with humanity the new master of creation. As the satirist Alexander Pope put it in his epitaph for Isaac Newton: 'Nature and nature's laws lay hid in night / God said, "Let Newton be!" and all was light.'[30]

Few thinkers embodied this belief more strongly than the Marquis de Condorcet, a mathematician, economist, and philosopher who argued that only the sciences could fulfil the promise of 'the indefinite perfectibility of mankind'. He divided the world into ten epochs, the eighth stretching 'from the invention of printing to the time when science and philosophy shook off the yoke of authority', the ninth ending with the 'founding of the French Republic', and the tenth – the most glorious yet to come – covering 'the future progress of the human mind.'[31]

Condorcet is a fascinating individual, both sincere and sympathetic. He is the last major French Enlightenment thinker and the only one to become seriously involved in the Revolution. He was a man whose combination of outward shyness and inner passion led contemporaries to describe him as both a 'snow-capped volcano' and an 'enraged sheep'.[32] He started the Revolution a monarchist, quickly came on board with the republicans, and died a democrat. But through his life he retained unwavering support for certain moral principles. He opposed slavery and the death penalty, for example, and advocated for equality of the genders and universal education. The core of his belief was that it was only by improving life for the common woman and man that humanity could achieve a new era of peace and harmony.

Condorcet would eventually be devoured along with so many of the Revolution's children. After spearheading a plan for constitutional reform put forward by the moderate Girondin faction, a warrant was issued for his arrest in 1793, when the radical Montagnards took power during the Reign of Terror. It was while hiding from his captors that the irrepressible Condorcet wrote what would be his masterpiece: the *Esquisse d'un tableau historique des progrès de l'esprit humain* ('Sketch for a Historical Picture of the Progress of the Human Mind'). The *Esquisse* was unfinished, a literal sketch, but is often seen as a final thesis of the Enlightenment; the apotheosis of its ideals, scribbled down by a hunted man and lobbed hopefully into the future.

It's confidence in generations to come that is the focus of the *Esquisse*, with Condorcet making an impassioned plea for a concept that is now commonplace: a belief in the accumulative virtue of humanity – that the world is growing in wisdom, and that life tomorrow will be better than it is today. In other words: a belief in Progress. As Condorcet writes: 'The time will therefore come when the sun shines only on free human beings who recognise no other master but their reason; when tyrants and slaves, priests and their benighted or hypocritical minions exist only in the history books and the theatre, and our only concern with them is to pity their victims and their dupes, maintain a useful vigilance motivated by horror at their excesses, and know how to recognise and stifle, by the weight of reason, the first seeds of superstition and tyranny that ever dare to reappear.'[33]

Condorcet justified this optimism by pointing to the successes of the scientific method. 'A young man leaving school today knows more mathematics than Newton acquired by profound study or discovered through his genius,' he notes. 'The same observation is applicable to all the sciences, though in unequal measure.'[34] He happily acknowledges that mistakes can be made, in both science and governance, but sees these only as precursors to future improvements. His is a philosophy of resilience and optimism. The only challenge is ensuring everyone has access to the fruits of progress. This means that individual thinkers should not only be unconstrained by the edicts of tyrants and priests, but that they should also have material support. The main barriers to improvement, says Condorcet, are the three inequalities: inequality of wealth, inequality of status, and inequality of instruction. Remedy these and society becomes a huge improving machine, the gears of education and teaching and learning working in a great virtuous cycle, each generation improving upon the last. '[A]s facts multiply, the human mind learns to classify them and reduce them to more general facts, and the instruments and methods used to observe and measure them acquire a new precision,' writes Condorcet. 'As the mind reaches more complicated combinations, simpler formulae make them easier to grasp.'[35]

In order for this cycle to establish itself, though, certain thinking tools are needed. Foremost among these is the development of a 'universal language': a tool that would allow the application of reason to all fields of enquiry with the clarity of mathematics. Such a language 'would serve to bring to all objects embraced by human intelligence a rigour and precision that would render knowledge of the truth easy and error almost impossible.'[36] Elsewhere, he imagines a decimal system that could be used to classify information. Each entity in the system would be described with a ten-digit code, with each number expressing a different permutation of a primary quality. So, an entry with the designation 4618073 might be located in the kingdom *animalia*, class *mammalia*, order *carnivora*, family *felidae*, species *F. catus*, with the colouring *tabby* and the body type *obese*. In other words: a fat tabby cat.[37] Condorcet imagines this system of universal data entry capturing every type of fact and statement imaginable. Such information could then be rendered in vast tables, like those advocated by Francis Bacon the century before, storing not just scientific knowledge but moral and political wisdom too.[38]

looking not just to fulfil abstract ideals but also to solve pressing concerns regarding the French economy. Condorcet himself was associated with a group of economists now known as the physiocrats, who argued that agricultural productivity was the cornerstone of national prosperity. As Karl Marx explained much later in *Das Kapital*, 'the Physiocrats insist that only agricultural labour is productive, since that alone, they say, yields a surplus-value'.[42] What was significant about their work, though, was their methodological approach, which stressed measurement and analysis as the right way to shape a nation's economy. Ruling by diktat was fine, but such orders had to be informed by reason.

In the eyes of the physiocrats, the measures of the Ancien Régime were undesirable primarily because they were an impediment to this reason. The old units were confusing and their elasticity imprecise. If the *aune* could change length from one piece of cloth to the next, what good was measuring it in the first place? The metre, on the other hand, would be 'the unit of an emergent property-owning democracy',[43] able to demarcate land accurately and consistently, define ownership, and improve taxation and agricultural yields. Metric was constant and universal: the perfect tool for measurement that took place on paper rather than in person. And in the same way that quantitative methods had given the likes of Galileo and Newton new mastery over the heavens, these new weights and measures would let France's elite better organise the fortunes of the nation.

Not coincidentally, the *savants* involved in the metric system's creation belonged to the class who would benefit most from such changes, while some members had an even more direct involvement. Condorcet was a former Master of the Royal Mint, while Lavoisier was a tax farmer – part of an elite group that helped collect the king's taxes (and kept part of the proceeds for themselves). The position earned Lavoisier 'one of France's great fortunes, as well as the hatred of millions of ordinary French men and women,'[44] as the practice of tax farming was much despised. It was a frequently mentioned grievance in the *Cahiers*, where the farmers were dubbed *sangsues de la Nation*: bloodsuckers of the nation.[45] Lavoisier's involvement in this practice would ultimately cost him his life when he was executed, along with twenty-seven other tax farmers, for alleged fraud in 1794, five years before the final metre was unveiled. His colleague on the Commission of Weights and Measures, Joseph-Louis Lagrange, mourned the loss, noting: 'It took them only an instant to cut off this head, and one hundred years might not suffice to reproduce its like.'[46]

From the standpoint of merchants, bureaucrats, and tax farmers, the adoption of reformed weights and measures would make business smoother. No longer would they have to deal with a farrago of uncoordinated units, of informal and unaccountable changes to the sizes of bushels and ells. Instead, everything would be unified, systematised, and *rational*. One politician involved in metrication, the former engineer Prieur de la Côte-d'Or, said the introduction of the new standards would make commerce in France 'direct, healthy, and rapid', moulding the country into 'a vast market, each part exchanging its surplus'.[47]

Metric reform would therefore be market reform, and a radical one at that. But the *savants* were able to push successfully for these changes because they stressed, time and time again, the naturalness and rationality of their new system. These concepts might seem contradictory, but were resolved by Enlightenment thinkers by framing science as a way to harness and control the natural world. On a political level, this allowed metric advocates to rally support; they could deride their opponents as irrational and unnatural, while defusing criticism that their measures were partisan by stating that they were taken from a neutral party: the Earth itself. Older measures were criticised as arbitrary and irrational, their use bolstered only by the tyranny of outdated political systems and the stupidity of the uneducated. Each of the old units had 'no other right to be a standard than [it has] certain marks upon it and a certain name given to it', they argued, while the metre and kilogram were 'perfect', 'true', and 'objective'.[48]

The inconsistencies in this argument can be seen whenever we dig into the practical work of actually defining and realising these new units. To realise the kilogram (originally known as the *grave*), for example, you had to weigh the mass of one cubic decimetre of water, a process that sounds natural enough. But scientists soon found that not just any water could be used for this task, as water sourced from the sea, lakes, or mountain streams varied in its composition, altering the kilogram's final weight. Only purified water could be used, which could only be sourced from a laboratory. Similarly, the temperature of this water had to be fixed using newly improved thermometers, while the weighing process had to take place in a vacuum, an environment as natural for the Earth as a forest on the moon. The contradictions embodied in this process of realisation reflect the broader contradictions of the metric system as a political project. These tensions would ultimately undo France's first attempt at metrication just years after it had begun.

The decimalisation of time

On 20 *Prairial*, Year II of the Revolution (8 June 1794), the flat fields of the Champ de Mars, a military drill ground near the centre of Paris, sprouted a mountain. It was no republican miracle, but it was intended to induce one. The mountain (more of a hillock, really) was constructed as the centrepiece of the Festival of the Supreme Being, a nationwide celebration orchestrated by Maximilien Robespierre to inaugurate his new civic religion, the Cult of the Supreme Being.

As with earlier revolutionary festivals, the event was stuffed and frilled with symbol and gesture. There were chariots carrying cornucopias of plenty, marching ranks of breastfeeding mothers, and choirs singing republican hymns. At the day's climax, Robespierre himself descended the mountain like Moses to deliver the new gospel: a deist and nationalist vision that presented the Supreme Being of Creation – the Author of Nature and in no way contiguous with the Christian God – as looking down with fatherly appreciation on the struggles of the French people, urging courage and fortitude. 'Is it not He whose immortal hand, engraving on the heart of man the code of justice and equality, has written there the death sentence of tyrants?' asked Robespierre,[49] who had spent the past year consolidating power and executing thousands of counter-revolutionaries in the name of virtuous terror. At the end of his speech, he set fire to a papier-mâché statue representing Hideous Atheism, which burned away to reveal another, flameproof statue contained within: Wisdom.[50]

The festival was a high-water mark in the revolutionary project; a moment where, to paraphrase the nineteenth-century German philosopher Hegel, man walked on his head and remade the World in accordance with Idea.[51] The ruling Jacobin clique had embraced the paradigms of rationality and nature championed by the *savants* and found in them justification for their own utopia. The festival was an instantiation of those ideas, as crudely didactic as burning a statue. The metric system and its egalitarian principles were happily incorporated into this new order, with metre and kilogram taking their place in a symbolic language of republicanism. Just as one could look republican and speak republican, wearing the *tricolore* cockade and addressing your neighbours as *citoyen* and *citoyenne*, using metric units showed allegiance to the cause. This was the remaking of the world.

The principle of decimalisation, now established as improving for the masses, would be a prominent element of this process.

A few weeks into the Terror, in October 1793, the deputies of the Convention voted to replace the current Gregorian calendar with a new Republican calendar, featuring twelve newly named months comprised of three ten-day weeks, or *décades*. Even the day itself was to be decimalised: divided into ten hours, with each hour split into a hundred minutes, and each minute comprised of a hundred seconds. This meant there would be a neat 100,000 seconds in a day (as opposed to 86,400 previously), which in turn necessitated a slightly shorter metric second, 0.864 of today's length. This was perhaps the ultimate expression of Idea shaping World, with one historian describing the republican as having 'seized revolutionary control of the very nature of time.'[52] It's an innovation Fritz Lang borrows in his 1927 film *Metropolis*, where the punishing work schedules of the underclass are set by a ten-hour clock.

The calendar was the more complete of these two creations, and the subject of much committee discussion. Early proposals for its ordering included a hyper-rational naming scheme ('first month', 'second day', and so on) and a version dedicated to revolutionary themes (it began with the month of *Régéneration* and ended with *Égalité*[53]). But the scheme that stuck was agricultural, intended to underscore not only the natural rationality of the new calendar, but also the regime's allegiance to labourers. The new months were renamed after either weather (*Nivôse* for the snow and *Pluviôse* for the rain) or work (*Vendémiaire* for the grape harvest and *Fructidor* for fruit), while days were dedicated to agricultural products. That meant crops, flowers, and vegetables for most of them, with an animal for the fifth day and a tool for the tenth. So in the second *décade* of *Prairial* (the month of meadows), the days are dedicated to the following: strawberry, woundwort, pea, acacia, quail, carnation, elderberry, poppy, linden, and pitchfork. In the words of the calendar's creator, poet and dramatist Fabre d'Églantine, such objects were 'much more precious, without doubt, in the eyes of reason than the beatified skeletons pulled from the catacombs of Rome'.[54] Not everyone shared this view, of course, and one contemporary writer in Britain mocked the calendar with his own set of names for the months: Wheezy, Sneezy, and Freezy; Slippy, Drippy, and Nippy; Showery, Flowery, and Bowery; and finally Hoppy, Croppy, and Poppy.[55]

Under the Republican calendar, the year was split into four seasons containing three months, with three weeks of ten days per month

AUTUMN

Vendémiaire	Month of grape harvest	September–October
Brumaire	Month of mist	October–November
Frimaire	Month of frost	November–December

WINTER

Nivôse	Month of snow	December–January
Pluviôse	Month of rain	January–February
Ventôse	Month of wind	February–March

SPRING

Germinal	Month of germination	March–April
Floréal	Month of flowering	April–May
Prairial	Month of meadows	May–June

SUMMER

Messidor	Month of harvest	June–July
Thermidor	Month of heat	July–August
Fructidor	Month of fruit	August–September

The imposition of the Republican calendar was not necessarily the bizarre or unreasonable act it's sometimes perceived as now. Public conception of time in the eighteenth century was pluralistic compared to today. Multiple calendars were layered over the year like sheets of tracing paper, each highlighting its own domain and interests. There was the Catholic Church's calendar with its parades, saints' days, and festivals; the almanacs of the farmers, with their emphasis on seasonal changes and work; and the legal and financial calendars of bureaucrats, scribes, and merchants. Neither was the idea of reshaping a calendar particularly alien. It was regularly discussed by intellectuals prior to the Revolution, not least because of the slow adoption of the Gregorian calendar over previous centuries.

(The previous Julian calendar had been abandoned after sliding out of sync with the seasons, a side effect of a miscalculation involving the length of the solar year.) Great Britain and its colonies had only made the switch to Gregorian in 1752, a change that scrubbed eleven days off the year, and other nations clung to Julian dates for centuries more. Turkey only abandoned them in 1917, and Russia in 1918.[56]

The idea of changing calendars, then, was not without precedent. What did cause trouble was the removal of the Church's influence from the structure of the year – the true goal of the republican reform. The new ten-day week meant there was no more Sunday service. Instead, the tenth day (*décadi*) was celebrated as a civil holiday. Saints' days and religious festivals were also removed, events that had been a central feature of communal life in French society, adding 'a little glory and beauty to an impoverished existence.'[57] The revolutionaries were well aware of these attractions, but wanted to claim such pleasures as a result of their own munificence. The Festival of the Supreme Being was one example, and there were countless others dedicated to various themes: to Youth, to Victory, to Old Age, to Spouses, to the Sovereignty of the People, and so on.

As the French historian and philosopher Mona Ozouf argues, such events went beyond simple propaganda to reorient space and time around new citizens. The year was now punctuated by republican virtues, while the physical openness of festivals represented the freedom and equality of the new France. 'The festival was an indispensable complement to the legislative system, for although the legislator makes the laws for the people, festivals make the people for the laws,' writes Ozouf. 'Through the festival the new social bond was to be made manifest, eternal, and untouchable.'[58]

Decimal time was undoubtedly part of this same programme, though it never carried anywhere near the same influence as the Republican calendar. Those who did use decimal time tended to be arch-rationals like Laplace or dedicated revolutionaries like the politician Louis Saint-Just (the right-hand man of Robespierre, he had a decimal pocket watch on him when he was arrested during the Thermidorian Reaction). There are records of decimal clocks being installed in public places but no evidence that this new timekeeping was embraced by the average citizen. Many of these clocks survive today and are wonderfully uncanny objects, their ten-hour faces and subtly altered seconds marking them out as visitors from another timeline.

The new days and dates were not totally rejected. 'Monsieur Dimanche' and 'Citoyen Décadi' became popular allegorical figures, with the former representing the traditions of the Ancien Régime and the latter the newly decimalised citizen.[59] But the older calendars continued to be used alongside republican ones, and 'for most of the years of its existence we can detect an awkwardness and even an embarrassment about the artificiality of the new system', according to historian Matthew Shaw. It was simultaneously a sign of change *and* of the impossibility of change – 'a constant reminder that the aims of the Revolution had not yet been achieved and an admission that the republic was, at best, a work in progress'.[60]

Decimal time was officially abandoned less than two years after its introduction, in the month of germination, 18 *Germinal* 1795, while the calendar, which found elite support but never mass adoption, continued until 1 January 1806. Eventually, it was condemned as tyrannical and misguided. Opponents said 'neither man nor beast' could work for ten days straight as the *décades* required,[61] while the seasonal descriptions of the months were a 'perpetual lie' outside of the south of France. The calendar, in other words, was unnatural and irrational, directly opposed to the *savants*' ideals.

Over time, the metric system began to attract similar criticisms. After it had been finalised in 1799, the government began the task of converting the population. They compiled conversion tables for metric units and each region's measures; printed pamphlets that expounded the system's virtues; and created standards of the metre and kilogram to be sent to each town. They deployed government agents to verify local standards and made metric units mandatory in the nation's schools. Undercover police, whose power grew under the increasingly authoritarian state, performed spot checks at markets. But it was not enough.

In the years after the unveiling of the metre, metrological confusion continued to reign in France. Shopkeepers and merchants kept multiple sets of standards behind their counter, a situation that encouraged exactly the sort of cheating and swindling the metric system was supposed to eliminate. When converting their prices into the new units, they rounded up their calculations, passing the cost of decimalisation on to buyers. Even government officials failed to keep up, with numerous examples of surveyors, accountants, and military engineers all persisting in their use of the old units. Alder notes one particularly telling incident in which the central office in charge of introducing the new units sent a set of metric standards to a local branch, with a receipt noting that the package weighed sixty *livres* in Ancien Régime units.[62]

Many perceived strengths of the metric system became liabilities on contact with the public. Defending the system's interconnected units, the Agency of Weights and Measures warned critics: 'You cannot attack a part of the system without endangering the whole. Otherwise many different objections will follow.'[63] But this unyielding stance led to ever stricter enforcement, undermining the principles of independence for the common citizen and smooth trading that metric units were supposed to foster. The government even set up offices that charged a fee to verify the weight of commercial transactions, a direct return to the feudal monopolies that had so incensed the common folk in the *Cahiers*.

The vogue for decimalisation even affected time, with a ten-hour day split into 100 minutes and 100 seconds, each second equal to 0.864 seconds today

As the political fervour of the Revolution dwindled, so did metric enthusiasm. By the time Napoleon seized power in the coup of 18 *Brumaire* 1799, concessions had already been made, with some metric units rebranded using Ancien Régime names. The decimetre became the *palme*, the centimetre the *doigt*, and so on. Two years after becoming emperor in 1804, Napoleon dumped the Republican calendar, restoring saints' days and Sunday worship in return for Church approval of his rule. Then, on 12 February 1812, what had become the First French Empire adopted the *mesures usuelles*, or 'ordinary measures', a new system that restored a limited selection of old units, like the *toise* and *livre*. Their value was still defined using metric standards, but units were no longer decimalised. Napoleon happily kept the consistency and accuracy of the metric project, but abandoned its more idealistic and troubling elements. 'Thus, after twenty years of trouble, mystery, and litigation, no advances are made, except that of having one common standard,' observed one British metrologist.[64]

Writing much later from exile on Saint Helena, after his attempt to unify the continent under his personal rule failed, the former Emperor Napoleon spoke his piece on the metric system. 'Nothing is more contrary to the organization of the mind, of the memory, and of the imagination,' he observed in his memoirs. 'The new system of weights and measures will be a stumbling block and the source of difficulty for several generations . . . It's just tormenting the people with trivia.' The would-be conqueror of Europe placed particular blame on the hubris of the *savants*, who, he claimed, had aimed too high and dreamed too much. 'It was not enough for them to make forty million people happy,' he said, 'they wanted to sign up the whole universe.'[65]

6

A GRID LAID ACROSS THE WORLD

_____

The surveying of land, the colonisation of the US,
and the power of abstraction

measures. He mixed with figures like the Marquis de Condorcet at intellectual salons, eagerly absorbed reports from the prestigious Académie des sciences, and followed the travails of the Revolution's political factions as they began the work of building a nation anew.

Unlike Condorcet, who spent his final years hunted by radicals, scribbling down half-finished plans for the betterment of mankind, Jefferson had the opportunity to implement his ideas on a grand scale. His workshop was the landscape of America, and his tool of choice a regimen of land surveys now known as the Public Land Survey System, or PLSS. It was this monumental project that divided much of the continent into 1-mile-square plots. This is the grid that hypnotises so many while flying over the US, but which first directed settlers and soldiers to expand the borders of the fledgling country, seizing the land from Native American nations who knew it so well. Theirs was the homeland that the grid was deployed to conquer.

Land surveys can seem like mere bureaucratic conveniences, but they play an important role in the development of the modern state. In his influential 1998 book *Seeing Like a State*, the political scientist James C. Scott argues that over the last few centuries, states have deployed various 'tools of legibility' to better understand and control the activities of their citizens. These tools are varied in both form and application, but share certain traits: they standardise and simplify the world, reshaping the organic development of society into forms that are more easily aggregated by administrative centres. Censuses are used to discover the size and composition of a populace, for example, and land surveys and property records document where they live and what they own. These methods of standardisation can touch on the most personal matters, reaching into the habits and customs of everyday life to adjust them for the benefit of unseen bureaucrats. Minority and regional languages are discriminated against or suppressed in favour of official languages, ensuring their speakers' assimilation into the dominant culture. And weights and measures of individual regions are replaced with standardised units that allow commerce to be similarly harmonised and surveilled.

Before such interventions could be carried out at scale, a government's scrutiny of its citizens was limited, says Scott. 'The premodern state was, in many crucial respects, particularly blind; it knew precious little about its subjects, their wealth, their landholdings and yields, their location, their very identity,' he writes. But these 'social simplifications' turned 'what was a social hieroglyph into a legible and administratively more convenient format'.[2] Such tools of legibility enhanced a state's power, inside and outside its borders. They allowed basic functions like taxation and conscription to operate more smoothly, while enabling entirely new forms of action, from public health initiatives and welfare programmes to political surveillance and suppression.

Take, for example, the introduction of surnames in Europe towards the end of the Middle Ages. Until at least the fourteenth century, writes Scott, the majority of Europeans did not have permanent patronymics, with individuals often adopting new names when starting a new job or moving to a new area.[3] This caused problems for the state when trying to track the activities of individuals, as illustrated by a court case from sixteenth-century England. Here, a Welshman is summoned to appear in court, but when asked for his name replies that he is 'Thomas Ap [son of] William, Ap Thomas, Ap Richard, Ap Hoel, Ap Evan Vaughan'. It's a perfectly normal name for the period, a genealogical title that is both intimate and informative, identifying not only the individual, but his ancestry. The information it contains makes sense to members of Thomas's community, who likely knew his father and grandfather before him. But to outsiders it is cryptic. The judge is unhappy and scolds Thomas, telling him to 'leave the old manner' and adopt a single surname that suits the administrative needs of the state. Whereupon Thomas Ap William Ap Thomas (etc.) 'called himself Moston, according to the name of his principal house, and left that name to his posteritie'.[4]

The courtroom christening of Thomas Moston underscores the driving purpose of tools of legibility: to iron out the particularities of local knowledge and repackage it into universal forms. Once you're aware of this dynamic, you will find it everywhere in your life, when the bureaucracies of state and business slot you into categories built for their convenience. We saw it also in the orderly cornfields of Iowa and, in contrast, the irregular pastures of Yorkshire. Although, in the latter case, England has certainly been subject to its fair share of top–down ordering (most notably through the process of enclosure), the country's long history has created a palimpsest of farmland, with new and old systems of land division overlapping and abutting one another. In Yorkshire, these divisions also obey the logic of the landscape, with boundaries fitted to the trajectories of hills and rivers. The plains of the Midwest, by comparison, are more accommodating to the planner, with the level landscape allowing Jefferson's survey, a grand social simplification, to express itself as a single, coherent system.

In the latter case, though, what was erased to create this pattern were the territorial claims of America's native populations. Writing in 1814, a few decades after Jefferson's land survey began, the Swiss–French intellectual and liberal politician Benjamin Constant observed that a new mode of domination had been unleashed upon the world, introduced by French revolutionaries and perfected by Napoleon Bonaparte. In the past, said Constant, 'the primitive conquerors were satisfied with outward submission; they did not inquire into the private lives or local customs of their victims'. But such 'local interests and traditions contain a germ of resistance', he notes, and so 'the conquerors of our times, peoples or princes, want their empire to possess a unified surface over which the arrogant eye of power can wander without encountering any inequality which hurts or limits its view. The same code of law, the same measure, the same rules, and if we could gradually get there, the same language.'[5] This is the mode of conquest that the survey enables: to capture the world in a single order. As Constant noted: 'the great slogan of the day is uniformity'.[6]

Beating the bounds

Nearly every agricultural society has developed some method for marking boundaries on the land. The ancient Egyptian rope-stretchers who restored the flood plains of the Nile to orderly farm-land were probably among the world's first surveyors (a word that comes from the Old French *sorveoir*, meaning 'to oversee'), while the Roman Empire developed a particularly robust method of land mea-surement known as the *centuriatio*. Using this system, the Romans divided territory across Europe into grids with a tool known as a *groma* – a tall staff with a horizontal cross on top and four plumb lines suspended from the ends, like a multi-occupancy gallows. The *groma* was used to sight straight lines and right angles, and even in antiquity played a crucial role in the operation of the state. For the Romans, the *centuriatio* not only simplified property rights and tax collection, but straightened roads for marching legions and portioned out farm-land gifted to retiring veterans. The survey, in other words, helped fund, direct, and reward Rome's imperial war machine.

Roman surveying is something of an historical exception, though, and older survey techniques more often relied on monumentation – the use of natural features and artificial markers to decide bound-aries. The importance of this practice is captured in ancient texts, with the Bible offering a number of warnings against disturbing such markers, as in Deuteronomy 27:17: 'Cursed be he that removeth his neighbour's landmark.'[7] As with Josephus's story of Cain introducing weights and measures, the need for the surveyor is seen as a sign of humanity's corruption and deceitfulness. In his *Metamorphoses*, Ovid describes our fall from a utopian Golden Age to our current parlous state, noting that 'the ground, which had hitherto been a common possession like the sunlight and the air, the careful surveyor now marked out with long-drawn boundary line'.[8]

The profession of land surveyor as we know it today, though, began developing in sixteenth-century Europe in response to various social trends, including a rapidly growing population vying for land.[9] The first printed treatise on the subject in English, the 1523 *Boke of Surveyeng*, barely mentions measurement, but instead focuses on estimations of acreage and verbal descriptions of 'metes and bounds' (metes being straight lines set between notable landmarks, while bounds are boundaries like forests, hills, streams, walls, and roads). In the decades that followed, this approximate art was replaced by the work of the 'land-meater', who calculated area using mathematical methods and geometric tools.[10] Although it was still common for land to be assessed using elastic values (how many mouths it fed, how long it took to plough, and so on), the new surveyor practised what was called 'platting' – drawing an estate as an illustrated map, or plat. The 1607 text *The Svrveiors Dialogve* ('Very profitable for all men to perufe', boasts the title page) assures readers anxious about these new-fangled methods that they can be quite as satisfying as traditional verbal descriptions: 'A plot rightly drawne by true information, describeth so the liuely image of a Mannor [that] the Lord sitting in his chayre, may see what he hath, where, and how it lyeth, and in whose vse and occupation euery particular is . . .'[11]

It's interesting to contrast this method with older traditions of survey and ownership in the British Isles, such as the annual ritual of 'beating the bounds'. During the beating of the bounds, residents of a town or village would gather together to carry out a foot survey of their community. Priests and elders would lead the expeditions, pointing out geographical features like streams, rocks, and walls that marked the limits of their parish. They would be followed by a gang of children armed with willow sticks, who would beat these objects to place them in their memory or, in earlier centuries, be subjected to hazing themselves to achieve the same end. One account of beating the bounds in Dorset tells how,

> if the boundary be a stream, one of the boys is tossed into it; if a broad ditch, the boys are offered money to jump over it, in which they, of course, fail and pitch into the mud, where they stick as firmly as if they had been rooted there for the season . . . if a wall, they are to have a race on top of it, when, in trying to pass each other, they fall over on one side, some descending perhaps into the still stygian waters of a ditch, and others thrusting 'the human face' divine into a bed of nettles.[12]

MAY. — Beating the Bounds.

In the era before accessible maps, communities in the British Isles would 'beat the bounds' to pass on geographic knowledge

This gruelling experience is only complete when the boys reach a sunny bank and are rewarded with 'a treat of beer and bread and cheese, and, perhaps, a glass of spirits'. It's the least one deserves after a faceful of nettles.

The practical motive for these rambunctious surveys was to pass down knowledge of parish boundaries from one generation to the next; essential work at a time when reliable maps were not readily available. But it's clear they also served a communitarian function, tying together the land and its people through ritual and memory. Accounts of beating the bounds note that the day could be used to settle disputes between neighbours and integrate new members into the community. They often ended on the village green with a communal feast shared by rich and poor alike, and had a distinct religious tone, taking place just before the feast of Ascension and accompanied by readings from the Bible. This is a form of surveying that not only defines the boundaries of a community, but strengthens the bonds between its occupants.

The beating of the bounds is still observed in a few places today, but dwindled in significance long ago as attitudes to land ownership changed. In Britain, this was due to the rise of cartographical surveying that accompanied the fall of the manorial system, a social and economic practice that emerged in Europe after the demise of the Roman Empire. Under manorialism, the lord of the manor, or *seigneur* (from the Old French *seignior*, meaning 'senior' or 'older'), holds power over the men and women of their estate. These commoners owe their lord rent in some form, but also hold certain rights, including access to communal land they can cultivate for their own benefit. Such land was usually divided into thin strips like the British furlong, with the boundaries worked out among the tenants. After the massive depopulation caused by the Black Death in the fourteenth century, though, manorialism declined. The labour shortage gave tenants new leverage to bargain for better rights, while sparse occupation meant merchants and rich farmers could buy land on the cheap. In this environment, the demand for maps and plats increased, not just as practical tools of survey but as badges of authority and ownership. Only someone with wealth and an active interest in the land would create a plat, which itself became a sort of deed – a record that could be integrated into local legal systems.

In England, land ownership also became tied to political sovereignty. When 1215's Magna Carta wrested certain feudal privileges away from the king, it gave freeholders of land new rights and protections. 'No Freeman shall be taken or imprisoned, or be disseised [dispossessed] of his Freehold, or Liberties, or free Customs, or be outlawed, or exiled, or any other wise destroyed,' reads a key clause, which remains on England's law books today.[13] By 1565, the Elizabethan lawyer Sir Thomas Smith was noting that land ownership had become the only way to guarantee one's participation in government: 'Day labourers, poor husbandmen, yea merchants or retailers which have no free land [...] have no voice nor authority in our commonwealth, and no account is made of them, but only to be ruled.'[14] And the following century, the philosopher John Locke made property one of the essential rights owed to all men by 'natural law', along with life and liberty. The right to ownership of the land could trump even a monarch's authority, thought Locke, so long as individuals followed certain restrictions. They should not take more than they need, leave enough for others, and work the territory to claim it, mixing sweat with the soil in a pact of ownership.

In such a world, measurement of the land was of the utmost importance. As a result, sixteenth-century England gave rise to one of the most widely used measuring tools in the world: the surveyor's chain, or Gunter's chain, named after its inventor, the seventeenth-century English priest and mathematician Edmund Gunter.

Gunter's chain is outdated technology, but a marvel nonetheless. First described in a 1623 textbook by its creator, each chain is made of metal, measuring 66 feet or 22 yards in length and divided into a hundred stiff metal bars or links, each 7.92 inches long. Metal made the chain more durable than the fraying and flexible ropes used before it, while the jointed construction meant it could be folded away for easy transportation. The real genius of Gunter's chain, though, was how it combined traditional English units of land measurement, based on multiples of four, with what was then the newest and most exciting mathematical innovation from the continent: decimals. The hundred-link chain is divided by brass tags every ten links, but the total length is also equal to four rods, a traditional measure of land that can supposedly be traced back to the length of a Roman military pike, or *pertica* (from which it gets its other common name: perch). This dual system meant a surveyor using the chain could measure in furlongs (equal to 10 chains, or 660 feet), miles (8 furlongs, or 80 chains), or acres (1 chain by 1 furlong, or 10 chains by 10 chains square). And if they wanted to measure in chains and convert to acres, they could simply divide their results by 10.

This flexibility made Gunter's chain the dominant tool for land surveying in the English-speaking world for some 300 years. And although it has long been superseded by modern measuring tools, Gunter's chain is still embedded in the landscapes of former British colonies, including the US, Canada, Australia, and New Zealand. Roads in these territories are often one chain wide, while building lots and city blocks are commonly measured in multiples of chains. In the UK itself, the length of the chain is encoded in one of the country's cultural cornerstones: the cricket pitch. It's proof that if you look hard enough at the divisions of the world that seem arbitrary or haphazard, you will find long-forgotten choices, produced by necessity and preserved by tradition.

Expanding West

These, then, were some of the legacies that Great Britain bequeathed to its former colonies: a belief in property rights as foundational to governance and the means by which to measure them out. And so in 1783, when the United States signed the Treaty of Paris that ended the Revolutionary War, it incorporated these concepts deep into its political superstructure. Inspired by Locke's ideas, many of the founding fathers believed that land would be America's saving grace: a 'solution to the fragility of republics', to quote historian Jeffrey Ostler.[15] Not only would the abundance of the American continent satisfy the material demands of the farmers, speculators, and slave owners who propelled its economy, it would also stop the development of the sort of small, despotic, land-owning class seen in England and France. However, for the land to do its job, the United States would need to claim it: to push out from its current crust of territory on the eastern seaboard and dive deep into the interior of the continent.

This work of expansion would be spearheaded by Thomas Jefferson, the son of a surveyor and a staunch believer in the political possibilities of dirt. In his only full-length book, *Notes on the State of Virginia*, Jefferson outlined his ideological vision for America's new republic. The text starts as a dry overview of the geography and resources of Jefferson's home state, but, like the lord of the manor admiring his plat, Jefferson is enthralled by the idea of the land and begins to rhapsodise over America's natural bounty and its potential to nurture a particular sort of liberty. The latter, he thought, would be achieved through the development of the land, an activity that Jefferson elevated to the status of religious virtue. 'Those who labour in the earth are the chosen people of God,' he writes, 'if ever he had a chosen people, whose breasts he has made his peculiar deposit for substantial and genuine virtue.'[16]

Europe could keep its dirty metropolises and crowded work-shops, thought Jefferson; America would till the soil, cultivating in its 'immensity of land' a vigorous, moral, and independent citizenry. These men – these *white* men – would own the land they stood on, making them impervious to bribery, debt, or other economic threats. 'The proportion which the aggregate of the other classes of citizens bears in any state to that of its husbandmen, is the propor-tion of its unsound to its healthy parts, and is a good-enough barom-eter whereby to measure its degree of corruption,' said Jefferson.[17] This view of land was shared by many other founding fathers. George Washington was a surveyor himself before he became a revolution-ary and looked to the frontier as 'the Land of promise, with milk & honey',[18] while Benjamin Franklin said the only 'honest way' for a nation to acquire wealth was through farming, 'wherein man receives a real increase of the seed thrown into the ground, in a kind of con-tinual miracle, wrought by the hand of God in his favor, as a reward for his innocent life and his virtuous industry'.[19]

The United States in these early years was a tumultuous place, a nation defined by shifting allegiances and uncertain borders. The Treaty of Paris may have established the limits of America's territory on paper, but sovereignty on the ground was contested, not only by the vestiges of European powers, but also by the indigenous Native American nations. The latter groups were not helpless bystanders, but communities that resisted and repudiated, accommodated and overlapped with colonial forces. '[W]hites could neither dictate to Indians nor ignore them,' writes historian Richard White. They 'needed Indians as allies, as partners in exchange, as sexual partners, as friendly neighbors.'[20] At the same time, the American state had to deal with the restless ambition of its own people: with settlers, spec-ulators, entrepreneurs, and politicians, who all strained against their nation's new borders. The French diplomat Alexis de Tocqueville wrote in *Democracy in America* that the sense of America's abundance overwhelmed and intoxicated its inhabitants. 'It would be difficult to describe the avidity with which the American rushes forward to secure this immense booty,' said de Tocqueville. 'Before him lies an immense continent and he urges onward as if time pressed and he was afraid of finding no room for his exertions.'[21]

Managing these competing interests was a priority for the federal government and it adopted a number of strategies to defuse conflicts between these groups. It bolstered the military, signed numerous treaties with Native Americans, and strengthened networks of communication and transport. The most important step taken by the federal government, though, was to claim authority over lands west of the first thirteen states. It would survey this territory, parcel it up into plots, and decide who could claim it and how. Land could be sold to speculators to clear debts from the Revolutionary War; gifted to veterans, just as the Roman Empire had; and traded with factions both indigenous and colonial. Controlling this resource was vital to the survival of the early American state, allowing it to set the 'direction, pace, and scale' of its expansion without straining its then limited economic, regulatory, and military capacities.[22]

Jefferson was critical in setting the manner of this expansion, and helped form a trio of laws passed in 1784, 1785, and 1787 and collectively known as the Northwest Ordinances. Along with the Declaration of Independence and US Constitution, these are among the most important documents in the founding of the United States – not because they contain impassioned cries for democracy, but because they describe, simply and plainly, how its land would be divided, sold, and governed.

The Northwest Ordinances authorised a survey in the form of a huge grid, initially covering the great mass of territory that lay between the founding states and the Mississippi River. The main subdivision of the grid was the 'township' – a 6-mile by 6-mile square (the decimally minded Jefferson had argued for 10 by 10) that contained thirty-six subdivisions of a square mile each.[23] The base unit of measurement would be Gunter's chain, 66 feet long, with its subdivisions of a hundred links. As a square mile contained 640 acres, the chain was a perfect unit for division, with each 40-acre plot (the smallest area of land that could be purchased) measured out in 20 chains' length. To measure the ground, a hindman would strike the end of the chain into the earth while the foreman walked ahead with the other end, obeying his partner's shouts to keep his footsteps dead ahead. In this way, thousands upon thousands of surveyors set out across America, each team moving, in the words of Scottish writer Andro Linklater, through forest and valley 'like a caterpillar, hunching up and stretching out, drawing a straight line to the west'.[24] It was an incredible work of cartographic self-invention: the wilful application of structure on to what was still a land of contested territory and uncertain borders.

The practical result of this patient work was that the federal government could control the expansion of the nation, directing the flow of settlers to specific areas as necessity demanded. It also framed this expansion in terms favourable to the settlers. The land prior to the survey was seen by the colonisers as unmanageable – wild and unsettled, or controlled by native groups. After the survey had passed over, chains jangling through the forests and scrub, it became pliant real estate that could be supervised at a distance and sold for as little as $1.25 an acre. As the French political scientist Émile Boutmy commented in 1891, when the surveyors had crossed the whole of the continent and the true scope of this land rush could be better appreciated: 'The striking and peculiar characteristic of American society is that it is not so much a democracy as a huge commercial company for the discovery, cultivation, and capitalization of its enormous territory. The United States are primarily a commercial society, and only secondarily a nation.'[25]

Sales were boosted by the simplicity of the survey.[26] A typical deed might be listed simply as 'Township 7 North, Range 4 West, Lot No. 20' – as easy to identify on the grid as a square on a chessboard and as simple to mark out in person. Settlers who roamed ahead of the official surveyors could even stake their own land by following the lines of earlier plots; as long as they connected with the great north–south meridians and east–west baselines that knitted the states together, they would more or less follow official measures.[27] Just as Condorcet had promised of decimalisation, Jefferson's grid made citizens 'self-sufficient in calculations related to their interests'.

Without complex or ambiguous descriptions of metes and bounds, the grid reduced arguments between settlers over boundaries.[28] The survey ensured 'perfect security of title and certainty of boundary, and consequently avoids those perplexing land disputes, the worst of all species of litigation', as one US senator wrote in 1832.[29] Purchases also tethered people to the authority of the federal government, without which their claims could not be verified. Each township included land for official institutions like universities and courts, whose presence meant early settlements would not lose 'all their habits of government, and allegiance to the United States'.[30] If the founding and early success of the United States were dependent upon the controlled spread of settlers across its plains, mountains, and rivers, then the settlers themselves were dependent on the guidance of the survey.

Although the original Northwest Ordinances only covered a portion of the continental US, the survey system would be reproduced in future legislation throughout the nineteenth century. As the United States expanded west through bloodshed, treaty, and commercial dealings (including the 1803 Louisiana Purchase from

An aerial photo of Nebraska taken from on board the International Space Station reveals the imprint of the US land survey

France of land to the west of the Mississippi, a cherished ambition of Jefferson's that almost doubled the size of the country overnight), the grid was repeated with methodical intent. Over the next two centuries, it would come to encompass more than 1.8 billion acres of land, covering three-quarters of the total area of the continental US, including thirty new states. There are still significant tracts that the survey has yet to reach, mostly in the depths of Alaska,[31] but to date, some 1.3 billion acres of rectangular plots have been sold off, with the rest kept for federal and state use.[32]

Just as Jefferson intended, the land grew citizens. 'The wit of man cannot devise a more solid basis for a free, durable, and well administered republic,' he wrote in a letter.[33] In the eyes of the settlers, the rigid, unwavering grid of lots and townships became the trellises on which the American polity would grow, with the right to own land a defining characteristic of this new people. To some observers, though, this seemed disturbingly egalitarian. When English writer Fanny Trollope, mother of the novelist Anthony Trollope, visited America, she noted that 'Any man's son may become the equal of any other man's son, and the consciousness of this is certainly a spur to exertion', but that this freedom was also 'a spur to that coarse familiarity, untempered by any shadow of respect, which is assumed by the grossest and lowest in their intercourse with the highest and most refined'.[34] For an arch Tory like Trollope, the abstractions of the grid had furnished Americans with a place to live at the expense of custom and community. 'No village bell ever summoned them to prayer [...] When they die, no spot sacred by ancient reverence will receive their bones [...] They pay neither taxes nor tithes; are never expected to pull off a hat, or make a courtesy; and will live and die without hearing or uttering the dreadful words "God Save the King."'[35] But for another British visitor in this same period, social theorist and sociologist Harriet Martineau, the allure of land offered a cultural stimulant that was, if anything, more powerful than Trollope's hallowed traditions. It provided psychological sustenance, giving citizens direction in the bewildering expanse of the New World. 'The possession of land is the aim of all actions, generally speaking, and the cure for all social evils,' wrote Martineau in 1837. 'If a man is disappointed in politics or love, he goes and buys land. If he disgraces himself, he betakes himself to a lot in the west ... If a citizen's neighbors rise above him in the towns, he betakes himself where he can be monarch of all he surveys. An artisan works, that he may die on land of his own.'[36]

This land is your land

But whose land was it? For when Jefferson started laying his grid across America, he was not revealing the bounds of some virgin continent, but expanding a process of dispossession begun by Spanish,

French, and English colonists centuries before. The surveyor's chain may not have been as directly responsible for the death and misery of indigenous people as the Winchester repeating rifle and smallpox virus were, but it was still an essential tool of colonial violence. The simplicity of the survey, the oversight and control it offered the federal government, and the psychological transformation it wrought in the minds of the people – strengthening their conception of the country as wild and unclaimed – all helped white settlers steal land from native tribes.

Colonisers often justified their acquisitions by arguing that the Indians wasted the resources that America offered. The flexible ecology of Native Americans, which balanced static farming with mobile activities like hunting and fishing, was derided as unproductive. 'They are not industrious,' complained one seventeenth-century colonist of Indians, 'neither have art, science, skill or faculty to use either the land or the commodities of it, but all spoils, rots, and is marred for want of manuring, gathering, ordering, etc.'[37] In 1808, Jefferson repeated these charges to a delegation of Native Americans in person, explaining to them the reasons for 'the increase of our numbers and the decrease of yours'. He argued that if their people could only imitate Americans, learning to 'cultivate the earth' and respect property rights, they could recover their strength. Ostler points out the irony of this particular speech, given that the Indian delegation Jefferson was addressing were Christian, literate, and knew how to farm. As Ostler notes, the president's inability to 'recognize that the Indians standing before him actually conformed to his criteria for civilization' reveals 'a powerful, though unarticulated, operating assumption of an innate racial inferiority'.[38]

The very concepts that Jefferson thought constituted civilisation were weapons to be used against native peoples. Treaties written in a language of mutual agreement were often signed at the end of a rifle, while the surveys of the federal government created the pretext for expulsion. As soon as indigenous people agreed to certain boundaries, they could be persecuted for broaching them. Treaties allocating territory to rival Indian nations were used to divide and conquer, with colonisers able to drive out one tribe without infringing on the 'rights' of the others, or pit the groups against one another.[39] Even without this trickery, the imposition of strict systems of land ownership and boundaries undermined the way of life for some Native Americans. Indians in the Great Plains region, for example, relied on nomadic movement across vast tracts of land to hunt, following the migratory paths of bison. But in tracing these routes, they would now find themselves unknowingly trespassing on territory claimed by colonisers – the surveyor's grid transformed into tripwires strung invisibly across their home.

Often, agreements between Native Americans and colonists were broken by settlers squatting the land illegally. Such activity could be officially disavowed by the government while serving its purposes. It created a foothold for further settlement and provoked Indian violence that could be met with military force. Surveying in this context gave the appearance of agreement between different groups, only for colonisers to later trespass the same boundaries. In his 1818 account of interactions between colonisers and Indian tribes, the missionary John Heckewelder records how, when Indians 'ceded lands to the white people, and boundary lines had been established [...] scarcely was the treaty signed, when white intruders again were settling and hunting on their lands!' If the natives complained, says Heckewelder, then the government would assure them the intruders would be removed, only to send in agents 'with chain and compass in their hands, taking surveys of the tracts of good land, which the intruders, from their knowledge of the country, had pointed out to them!'[40]

In the fertile Mississippi Valley, the expulsion of the Choctaw,

Creek, and Chickasaw peoples was quickly followed by a land survey in 1831. This survey laid the foundations for 'the greatest economic boom in the history of the United States to that point'[41] – the revitalisation of the country's cotton trade through slave labour. Cotton had been grown in America prior to the nineteenth century, but it was this era that saw the rise of King Cotton: the vast and sprawling plantation system that dominated the economy of the South. Estimates suggest as many as one million slaves were transported to work the fields in its heartland surrounding the Mississippi Valley between the years 1820 and 1860. The slaves were themselves subject to the same commercial abstractions as the land, sorted by traders into grades based on their physical fitness: 'Extra Men, No. 1 Men, Second Rate or Ordinary Men, Extra Girls, No. 1 Girls, Second Rate or Ordinary Girls.' As with the sale of plots of land, this categorisation hastened the sale of their bodies.[42] The ease with which the survey divided land into plots also enabled wealthy speculators to invest in the area. They saw the potential of the valley's rich soil and bought huge plots to better exploit the efficiencies of slave labour. Contrary to Jefferson's propaganda, his grid supported not just a prosperous yeoman citizenry, but also an enslaved and immiserated society: men, women, and children in chains, who mixed their blood and sweat with the soil, as Locke had envisioned, without expectation of ever claiming ownership.

The intrusion of settlers on American land is particularly significant given that so many aspects of Native American religion and culture are tied to an understanding of place as sacred. For many indigenous people, the land is not simply a resource but a living entity: a nexus of complex relationships between humans, animals, and spirits. Certain religious ceremonies can only take place in specific locations, and the landscape itself is a physical history of mythical and historical events. 'For Indian men and women, the past lies embedded in features of the earth – in canyons and lakes, mountains and arroyos, rocks and vacant fields,' writes anthropologist Keith Basso. Together, these overlapping spheres of meaning 'endow their lands with multiple forms of significance that reach into their lives and shape the ways they think.'[43] Put more directly in the words of Vine Deloria Jr., an American Indian author and activist: 'American Indians hold their lands – places – as having the highest possible meaning, and all their statements are made with this reference point in mind.'[44] Such meanings can't be mapped with a surveyor's chain.

Although there is significant contrast between the conceptions of land expressed by colonisers and Indians, it is also possible to over-emphasise these differences. To do so is to fall into the same patterns of thought as the colonisers themselves and to divide the world between the 'mapped' and the 'unmapped'. The real distinctions are not so clear. Native tribes often gained meaningful concessions from government treaties, and surveys could be used for their temporary benefit too. Similarly, settlers and federal agents embraced flexible attitudes to land when it suited them, claiming access to specific resources, like timber and water, rather than entire territories. It's not wrong to contrast the two worlds, but it can suggest that there was some insurmountable gulf of understanding between American colonisers and native tribes that made the process of indigenous destruction somehow natural or inescapable – just the wheels of the world turning. But this is not the case. The expulsion of American Indians from their land was not the sad result of incompatible world views or the 'natural' triumph of superior knowledge; it was a deliberate effort to steal land and found a white nation at the expense of millions of lives. As de Tocqueville observed with brutal clarity, the legal charades of property rights, treaties, and surveys allowed the United States to remove Indians from the land 'with wonderful ease, quietly, legally, and philanthropically, without spilling blood covered nearly 8,400,000 acres[48] of territory and is a milestone in cartography: the most detailed, accurate, and extensive cadastral survey of the early modern era, and the first conducted at a national scale. For those who commissioned it, the survey was a huge success, facilitating the transfer of land and control of the nation. After it was completed, Catholic land ownership in Ireland fell from around 60 per cent to 14 per cent,[49] resulting in 'the most epic and monumental transformation of Irish life, property and landscape that the island has ever known.'[50] It was a forceable change in the country's ruling elite that wouldn't be reversed for centuries, and, again, it was the power of the survey and measurement that was instrumental.

Similar stories can be told around the world in this period, from the hacienda system introduced by the Spanish empire in South America to the colonisation of New Zealand and Australia, and the so-called Scramble for Africa at the end of the nineteenth century. Maps made such conquest possible, both as practical tools to locate resources and coordinate troop movements, and as imaginative props that let settlers envision themselves as the first to set foot in unclaimed lands and as missionaries of modernity. Writing about the colossal Survey of India that preceded its incorporation into the British Empire, the historian Matthew H. Edney describes how the maps produced in this period didn't just offer support to the imperial project in the form of 'territorial integrity', but came to define its very existence. 'The empire exists because it can be mapped; the meaning of empire is inscribed into each map,' he writes.[51]

The modern survey, then, is in some ways a product of a very specific mode of colonial expansion, but the vantage points created by these tools still shape our conception of the world today. Political theorist Hannah Arendt described the work of surveying and mapping that began with the colonisation of America as one of three great events that 'stand at the threshold of the modern age and determine its character' (the other two being the Reformation of the Catholic Church and the cosmological revolution begun by Galileo). Before railways, steamships, and aeroplanes shrank the dimensions of our planet, says Arendt, there was an 'infinitely greater and more effective shrinkage which comes about through the surveying capacity of the human mind'.[52] Through this work we've condensed the immensity of the Earth into something easily seen and understood; that is the power of these systems. But our ability to grasp such information requires that we disentangle ourselves 'from all involvement in and concern with the close at hand'. This leads to what Arendt calls 'world alienation'. Instead of living in a realm constructed from overlapping forms of human experience and action, we operate inside an empire of numbers, symbols, and models, and, as a result, lose connection to both one another and our sense of self-identity.

I would qualify this a little and argue that the distanced viewpoints Arendt says lead to alienation can also create new connections that reach beyond the bounds of the 'natural' self to encompass more expansive communities. In the same way that our understanding of temperature has evolved, creating scientific mythologies to replace the ancients' conception of fire as a primal force, the shrinking of the world through cartographical means doesn't have to leave us impoverished. In the age of climate change, for example, our awareness of the world as a huge, interconnected system is a vital tool to motivate political action. It reminds us that the wildfires and floods afflicting one country are only symptoms of a disease that infects every nation on the planet. The modern age of alienation that Arendt describes has created new holistic visions, like James Lovelock's Gaia hypothesis, which argues that the world is a single living organism that strives to support complex life, or the overview effect, a phenomenon reported by astronauts, in which viewing the Earth from space suddenly impresses upon them the interconnected nature of all life. The latter was most memorably expressed by Edgar Mitchell, a member of the Apollo 14 team and the sixth person to walk on the moon. As he told *People* magazine in 1974: 'You develop an instant global consciousness, a people orientation, an intense dissatisfaction with the state of the world, and a compulsion to do something about it. From out there on the moon, international politics look so petty. You want to grab a politician by the scruff of the neck and drag him a quarter of a million miles out and say, "Look at that, you son of a bitch."'[53]

In the context of the history of measurement, I think what Arendt is describing is an example of a recurring tension between the particular and the universal; between knowledge that is tied to a specific time or place and information that has been abstracted from its source. We can see this same dynamic in the invention of the metric system, creating weights and measures that are equally valid in every territory, and in the history of the American grid survey too. Here, it's expressed in the friction between indigenous peoples' understanding of a spiritual landscape, which locates memory and history in specific places, and the commercial abstractions of the grid, which divorced the land from its past, creating plots that could be bought and sold on paper. The latter framework is not inherently malicious, but when animated by an ideology of racial hierarchy and white supremacy, as it was in the US, it becomes a powerfully destructive force.

Negotiating the abstractions of the grid and its many cousins is a duty we owe to ourselves and to others. In describing the effects of the state's tools of legibility on its people, Scott notes that 'categories that may have begun as the artificial inventions of cadastral surveyors, census takers, judges, or police officers can end by becoming categories that organise people's daily experience precisely because they are embedded in state-created institutions that structure that experience.'[54] It is a trap we must be wary of in deploying measurement and quantification. We've likely all experienced moments when such categories of convenience have run roughshod over our own lives; when the impositions of bureaucracy trample on personal experience. It might be a frustrating visit to a doctor when our symptoms don't fit the official diagnosis, or a test at school that rewards only pre-ordained forms of knowledge. Sometimes these moments are tragedies, as with the plight of stateless refugees, for whom a lack of legal status means restricted access to aid. For those trapped by such circumstances, the event can feel unreal, a version of their lives dictated on paper. But as Scott notes, it is 'on behalf of such pieces of paper that police and army are deployed'.[55] When the map is drawn, we cannot forget the land it only partially describes.

7

MEASURING LIFE AND DEATH

———————

The invention of statistics and the birth of average

I am afraid that the mathematicians, who have not yet trouble the world, will trouble it at last, and that their turn has come.

—LOUIS-SÉBASTIEN MERCIER, *THE NEW PARIS*, 1800[1]

The Bills of Mortality

A year after its founding in 1660, the Royal Society of London, now the oldest national scientific institute in the world, admitted into its ranks a lowly shopkeeper. The appearance of haberdasher John Graunt among the Society's gentleman philosophers seems to have caused a bit of a stir, but Graunt's application had been approved by no higher authority than King Charles II himself, the Society's patron. As one contemporary observer noted, the king made it clear that 'it was so farr from being a prejudice, that [Graunt] was a Shop-keeper of London; that His Majesty gave this particular charge to His Society, that if they found any more such Tradesmen, they should be sure to admit them all, without any more ado.'[2] It was praise from the highest station of the land, but how had Graunt earned it? Through his publication of a single, short book: *Natural and Political Observations Made Upon the Bills of Mortality*. It would be the only work Graunt would ever write, but is now recognised as a founding text of statistics: a discipline concerned not with individual measures but with the power of measurement in aggregate.

Graunt's work heralds the start of another revolution in metrology, one that expands the domain of measure yet again to encompass not just individual events and actors but complex entities comprised of either interlocking or independent parts. In other words: a form of measurement capable of holding a ruler up to society at large.

Graunt's own interests lay in the fates of his fellow Londoners. In his *Natural and Political Observations*, he compiled the weekly tables of births, christenings, and deaths (including their causes) published by London's parishes and known collectively as the Bills of Mortality. To these, he applied 'the Mathematiques of my Shop-Arithmetique'[3] in order to extract 'some Truths, and not commonly-believed Opinions'.[4] His stock-keeping of London life allowed him to compute a number of previously unknown facts, including the first reliable population estimates for London and the nation; the era's incredibly high infant mortality rate (which peaked in the capital but was balanced by rural migration); and numerous public health trends, including the rise of rickets, the under-reporting of syphilis, and the impact of 'that extraordinary and grand Casualty the Plague' on London's population. In the words of one modern epidemiologist, Graunt contributed 'more to human knowledge than most of us can reasonably aspire to in a full career'.[5]

It's easy to see why this work earned him the support of King Charles II. In a time before the state had developed various tools of legibility, insights like Graunt's were invaluable. No spy or diplomat could collect such intelligence, interrogating the whole country using just a little calculation and data collection. Indeed, Graunt's work even debunked a then popular theory that plagues struck with particular barbarity during the first years of a monarch's reign – a welcome absolution for the sovereign.

Graunt's analysis is all the more impressive given that the information he collected was already public. As he notes in the book's preface, the Bills of Mortality were read regularly by Londoners, but most 'made little other use of them, then to look at the foot, how the Burials increased or decreased'.[6] Why Graunt decided to compile

The Diseases and Casualties this Week.

French-pox	1
Griping in the Guts	34
Jaundies	1
Imposthume	3
Infants	11
Kild at St. Martins in the fields	1
Kingsevil	1
Livergrown	1
Palsie	1
Plague	267

Abortive	3
Aged	19
Bleeding	1
Cancer	1
Childbed	7
Chrisomes	14
Consumption	83
Convulsion	31
Dropsie	22
Drowned 3, two at St.Katharines Tower, and one at St. Magdalen Bermonsey	3
Executed	3
Feaver	48
Fistula	1
Flox and Small-pox	23
Flux	1
Found dead in the street at St.Peters in Cheapside	1

Rickets	9
Rising of the Lights	4
Scurvy	1
Shot with a Pistol at Saviours Southwark	1
Spotted Feaver	12
Stilborn	11
Stopping of the stomach	3
Strangury	1
Surfeit	14
Teeth	32
Thrush	3
Timpany	1
Tissick	2
Vomiting	3
Winde	3
Wormes	4

Christned	Males	107	Buried	Males	331	Plague	267
	Females	92		Females	353		
	In all	199		In all	684		

Increased in the Burials this Week————— 69
Parishes clear of the Plague———— 110 Parishes Infected——— 20

The Assize of Bread set forth by Order of the Lord Maior and Court of Aldermen, A penny Wheaten Loaf to contain Nine Ounces and a half, and three half-penny White Loaves the like weight.

John Graunt is considered one of the first statisticians, compiling his data from London's 'bills of mortality' decades' worth of figures isn't clear, but he was evidently aware that his work was unprecedented. He takes the time to include in his book his tables of calculation so others might check his workings, and goes out of his way to give others the opportunity to correct him. He writes of his proofs: 'For herein I have, like a silly Schole-boy, coming to say my Lesson to the World (that Peevish, and Tetchie Master) brought a bundle of Rods wherewith to be whipt, for every mistake I have committed.'[7]

Work like Graunt's would become more common from the seventeenth century onwards, as increasingly centralised states sought to better understand their own people. Graunt himself was a friend of the surveyor of Ireland, William Petty, who coined the term 'political arithmetic' to describe such numerical reckoning on a national level. Writing in 1687, Petty praised the capacity of such figures to capture the fates of nations: 'When I find out puzling and preplext Matters, that may be brought to Terms of Number, Weight and Measure, and consequently be made demonstrable . . . I willingly ingage upon such Undertakings.'[8] What Gunter's chain and the surveyor's plat did for land, political arithmetic would do for its people. It would quantify their lives: tabulating births and deaths, marriages and murders, morality and mortality.

The original object of such work was to steer government policy. As Petty wrote: 'To practice upon the Politick, without knowing the Symmetry, Fabrick, and Proportion of it, is as casual as the practice of Old-women and Empyricks'[9] (the latter being a term then synonymous with charlatans and quacks, showing that empiricism had yet to attain its current reputation). But the analytical tools that would develop within political arithmetic, many of which were borrowed from other scientific disciplines, would prove invaluable in other fields, and were eventually separated from their original object of study to become their own branch of knowledge. Graunt himself was aware of the potential of his work to address other questions, noting that his conclusions 'have fallen out to be both Political and Natural, some concerning Trade and Government, others concerning the Air, Seasons, Fruitfulness, Health, Disease, Longevity, and the proportions between the Sex, and Ages of Mankinde'.[10] From the start, it was clear that the analysis of aggregated data was useful in fields we would now identify as biology, meteorology, and epidemiology. And why these particularly? Because they concern phenomena where single observations and measurements can only explain so much. If you want to understand how a disease spreads in a population, you can't study a handful of patients and extrapolate from there. If you want to know what the weather is like in your country, simply looking out of the window and assuming it's raining *everywhere* won't help. You need to gather together individual measures and then examine them as a group. Only then do you have a frame of reference equal to the scope and variety of your subject and can find new patterns and trends. This is why statistics are so incredibly useful. In contrast to the first units of measure, literally derived from the body, statistics operate at a scale far beyond the grasp of the individual.

This is perhaps why statistical measures have become commonplace in the modern world, where so much of life seems interconnected. Despite this, their status often goes under-examined, the method of their making overlooked. Discussions of unemployment rates, inflation, and population growth are ordinary enough, yet all these figures 'hover between the realms of the invented and the discovered', as historian of science Lorraine Daston puts it.[11] They are robust enough to set government policy yet often fall apart when subjected to scrutiny. Take gross domestic product, or GDP, for example, a statistic widely accepted as shorthand for a nation's economic health. It's used to decide spending priorities around the world, and if a country's GDP falls unexpectedly, then think tanks will thunder, of the plague on London, would likely have been surprised by the ubiquity and power of these figures, which not only tally death, but govern life.

The combination of observations

The origins of statistics can be traced back, like so many things in metrology, to observations of the night sky. It was this celestial black-board that offered humanity its first lessons on the importance of mensuration, teaching astronomer-priests how to predict the sea-sons and so laying the foundations of quantified science. It's no sur-prise, then, that it was an astronomer, the Belgian Adolphe Quetelet, who first brought the logic of quantification to bear on human lives. Compared to some of the thinkers mentioned so far, Quetelet's work is mathematically limited, but his belief in the universal applicability of statistics was striking and prodigious. Like the Oxford Calculators before him, Quetelet introduced measurement to new arenas, show-ing that it could reveal not just the workings of inanimate phenomena but, as he supposed, innate truths about society. Today, such figures steer the work of nations as persuasively as the stars did our ancestors.

Quetelet was born in the city of Ghent in 1796, and was, by most accounts, a man of varied interests. Academically adept and artisti-cally minded, he published poetry, performed in plays, and at one point dreamed of becoming either a painter or a sculptor.[15] At the age of twenty, he discovered an aptitude for mathematics that earned him the first doctorate ever awarded by the city's university, and his work was significant enough that the Dutch government (then in charge of present-day Belgium) summoned him to Brussels to teach mathematics. There he became involved in a project to build a new observatory and was sent by the government in 1823 to Paris, the scientific capital of the age, to learn from the same *savants* who had constructed the metric system. Many of those individuals continued to live and work in the city, including the polymath Pierre-Simon Laplace and mathematician Adrien-Marie Legendre. It was from this group (though not necessarily under their direct tuition) that Quetelet would acquire the skills he needed to run an observatory back home, learning, most importantly, how to catalogue and analyse data.

Astronomers at this time had begun to re-evaluate their relationship with measurement through the concept of error. Mistakes in scientific observations have always been expected, but scientists have not always known what to do with them. When observing events in the night sky, for example, an astronomer might take multiple readings but not worry too much if they varied. Some individuals might try to combine their measurements, but most would just select a single 'golden number' they thought best represented their work.[16] Typical of this attitude was the advice of the Anglo-Irish chemist Robert Boyle, who, in 1661, noted that 'experiments ought to be estimated by their value, not their number [...] As one of those large and orient Pearls that are fit to adorn a Monarchs Crown, may outvalue a very great number of those little (though true) pearls that are to be bought by the ounce in Goldsmiths and Apothecaries shops.'[17] Irregular results were not deemed profitable, as they are today, able to indicate unexpected findings or hone experimental methods. Instead, they were unwanted and even shameful offspring, evidence of a lapse in focus or lack of skill.

This dynamic was felt keenly in astronomy, where improvements in measuring tools – particularly the telescope – led to an increase in errors. For accuracy to create mistakes may seem paradoxical at first, but think of it like this: if you have to measure your height twenty times in quick succession, the first ten times using a tape measure marked in feet and inches, and the second ten using a laser that judges length to the millimetre, which set of results will show more consistency? It's easy enough to hit the mark of 5 foot 10 inches ten times in a row, but measuring out 1.778 metres over and over again is a bigger challenge. This was the problem that afflicted astronomers: their tools were more precise, yes, but that precision meant they responded to factors that were difficult or even impossible to control. The sensitivity of telescopes registered not only the location of celestial objects, but distortions caused by weather, inconsistencies in their lenses, and regular human error. Indeed, this is one of the fundamental traps of measurement: the more precise you are, the more inconsistent your results often appear to be.

Acclimatising to this frightening, error-strewn landscape was a slow process. An early breakthrough was made by eighteenth-century German astronomer Tobias Mayer, who was fixated with measuring the librations of the moon – wobbles on its axis caused by eccentricities in its orbit and inclination. For each measurement Mayer took, he made a trio of recordings at set times during the day, before creating an average of their values using simple equations. This approach seems so straightforward 'that a twentieth-century reader might arrive at the very mistaken opinion that the procedure was not remarkable at all', notes historian of statistics Stephen M. Stigler.[18] But combining data in this way was, at the time, extremely unusual. When Mayer's contemporary, the renowned Swiss mathematician Leonhard Euler, was faced with a similar problem, he concluded that if observations were combined, 'error has then multiplied itself'.[19] But Mayer made the conceptual leap that errors in measurement don't have to compound and can instead cancel one another out. He dubbed his approach 'the combination of observations', a phrase that was used to refer to statistical methods in general throughout the nineteenth century,[20] and which, at the time, must have seemed like mathematical alchemy: transmuting error into accuracy.

Another breakthrough was the discovery of the error law, otherwise known as the normal distribution, or bell curve. This is an entity of mythic dimensions in the history of mathematics, with a complex past and many variations. Most simply, though, it can be described as a pattern of error. The underlying premise now seems like common sense: that when we repeatedly take measurements of something, our results tend to cluster around a single value representing the 'true' figure, accompanied by a large number of readings close to this central point that dwindle in number the further we stray. This phenomenon had been known to astronomers for a long time, with Galileo himself commenting on it, but was not explored systematically until the eighteenth century. It was a French mathematician, Abraham de Moivre, who was first to offer an analysis, when investigating problems of probability in gambling. Imagine, for example, you want to know what number is most likely to come up when rolling a pair of dice. You know that there are thirty-six possible combinations and eleven different outcomes (two through twelve). By adding up how many ways each total can be reached using the two dice, you'll find the likeliest result – in this case, seven – that can be produced by six different combinations. Plot those figures as a bar chart and you'll notice that the resulting diagram forms a familiar shape: a symmetrical curve that humps in the middle and tapers off at either side. This is the normal distribution.

What gives this figure its grandeur, though, is its ability to refine data and minimise error. De Moivre's handling of the normal distribution was relatively limited (he was primarily interested in what are called binomial distributions), but later thinkers realised the full potential of the concept as a way to describe probability. Two individuals in particular, the astronomer Laplace and the German mathematician Carl Friedrich Gauss, rendered the normal distribution and its many variations in generalised mathematical form, turning the resulting equations into crucibles that could melt away the impurities from cluttered tables of error-strewn observations, leaving behind what seemed to be true and untarnished results. After all, are scientific observations really so different from the results of a game of chance? In both cases there are desired outcomes and variables that influence them. In backgammon, it might be the roll of a dice; in astronomy, it's the breeze that nudges your telescope. Working out exactly which of your results is the most unlikely can help rationalise this chaos. Along with a handful of connected mathematical concepts (most notably, the method of least squares and the central limit theorem), the normal distribution not only revolutionised how scientists conceptualised error, but helped them marshal ever larger and busier tracts of data, laying the mathematical foundations for the discipline of statistics.

The birth of the average man

Quetelet had gone to Paris to observe the night sky, but had his head turned by constellations of data instead. On his return to Brussels in 1824, as well as taking over the city's observatory, he was tasked with assisting a government census. The sight of these administrative records, so similar in format to celestial observations, confirmed in him an idea first suggested in France. If astronomers could use careful measurement to reveal not just facts about the cosmos but laws that governed its movement, why couldn't Quetelet do the same for society? He believed that anyone who spends long enough thinking about the universe will be 'struck with the admirable harmony that reigns there [and] can not be persuaded that similar laws do not exist in the animate world'.[21] This analogy would be the founding ethos of his new science: a discipline he dubbed social physics.

Over the subsequent years, Quetelet sought to apply the statistical tools developed in astronomy to the data of political arithmetic. It was the perfect moment for such work, with nations in this period collecting more data about their citizens than ever before. There was 'an avalanche of printed numbers', in the words of Canadian philosopher and historian Ian Hacking, as states 'classified, counted and tabulated their subjects anew'.[22] To give some idea of this growth, consider that the first ever US census of 1790 asked each household just four questions: how many men, women, other free persons, and slaves live here? The 1880 edition, by comparison, encompassed some 13,010 queries, directed not only to individual households, but also to farms, factories, hospitals, churches, universities, and every other sort of institution that might render this newly vital resource to the government.[23]

In 1835, Quetelet published a two-volume work titled *Sur l'homme et le developpement de ses facultés*, translated into English in 1842 as *A Treatise on Man and the Development of His Faculties*. In one typical analysis from the book, he sorts through records of chest measurements belonging to exactly 5,738 Scottish soldiers, looking for patterns in the data. He shows that if these figures are plotted as a bar chart, they reveal the same pattern de Moivre discovered in games of chance: the normal distribution, or 'law of error', as it was then known. Just as scientists had before him, Quetelet reasoned that the middle value of this curve was the 'true' figure, which he took to be the outcome favoured by the laws of nature. He collated similar data sets for height, weight, and the size of various body parts, cataloguing them by nationality. He even invented new forms of measurement, including the body mass index (known as the Quetelet index for many years), which he derived from the ratio of weight to height. By collecting this data and fitting it like a statistical Frankenstein into a single body, Quetelet introduced to the world stage a character never seen before: *l'homme moyen*, or the average man – an amalgamation of statistical regularities whom he thought embodied the typical attributes of each nation.

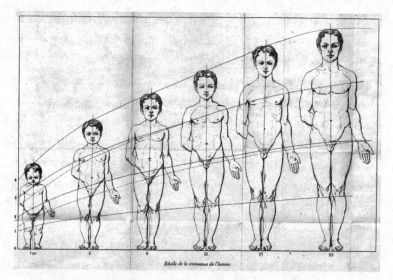

Echelle de la croissance de l'homme.

Belgian statistician Adolphe Quetelet applied numerical techniques borrowed
from astronomers to new indices of data, including the human body

In our time, the concept of the average is most likely to be seen
as an insult, as an indictment of mediocrity. But for Quetelet, to be
average was to be perfect. He saw the average statistics in his biometric data as the proper outcome of natural laws; an indication that
the subject was free of all abnormality. The average man, he wrote,
exists 'in a just state of equilibrium, in a perfect harmony, equally
distant from excesses and defects of every kind, in such a way that . . .
one must consider him as the type of all that is beautiful and all that
is good'.[24] As Hacking notes, Quetelet's achievement in *Sur l'homme*
and his subsequent works is a subtle but significant transformation of
what statistical measures mean.[25] When the normal distribution was
applied in astronomy, the average reading was not necessarily taken
to be the real value, but a best approximation accounting for random error. Quetelet, however, transformed this mathematical shorthand into a 'real quality' – something that was true of the groups of

humans he studied. It is an ontological reordering that turns abstract measure into something tangible, and while the result is not necessarily false, it is definitely constructed, and therefore fallible.

Quetelet's work was received with fascination and acclaim. A three-part review in London's *Athenaeum* concluded: 'We consider the appearance of these volumes as forming an epoch in the literary history of civilization.'[26] His work triggered a statistical feeding frenzy among the professional classes, with societies, journals, and institutions devoted to the discipline springing up across Europe. Members collected and published data of increasingly dubious relevance, with one resourceful acolyte siphoning off the contents of the toilets in a busy train station to attempt to deduce the 'average European urine'.[27] And Quetelet's own work began to expand in scope too, as he looked beyond the dimensions of the human body to the wider patterns of societal data. Here he distinguished himself not as a mathematician but as a miner of quantitative ore, delving into government publications and scientific journals to find data to work with. He collected statistics on crime and suicide rates, on marriages, births, and disease, and cross-referenced his findings by age, sex, occupation, and place of residence.

The more Quetelet collected, the more he discovered shocking regularity in society's workings. Even the most passionate activities – marriage and murder, for example – seemed to barely differ year by year in many countries. The results disturbed and excited him in equal measure. In one analysis of Parisian police statistics, he notes 'the terrifying exactitude with which crimes reproduce themselves' – not just in the number of murders each year but in the proportion of stabbings to shootings to strangulations. 'We know in advance how many individuals will dirty their hands with the blood of others, how many will be forgers, how many poisoners, nearly as well as one can enumerate in advance the births and deaths that must take place,' he wrote. Such figures constituted a 'kind of budget for the scaffold, the galleys and the prisons, achieved by the French nation with greater regularity, without doubt, than the financial budget'.[28] The consistency of these findings revealed to Quetelet that individual choice mattered far less than we had ever thought. You might believe you're getting married out of love, but statistics show you are simply following a line on a graph. As Quetelet put it in a private letter to a friend: 'It is society that prepares the crime; the guilty person is only the instrument who executes it.'[29]

In comparison to his earlier work, these new conclusions triggered feelings of anger and disbelief. They sparked debates about the nature of free will and the possibility of human progress in the face of statistical regularities. 'The great reviews and literary magazines were deluged by discussions of fatalism,' writes Hacking. 'No topic was more intensely discussed before it faded into oblivion.'[30] Some of the most illustrious figures of the nineteenth century were influenced by the discoveries of Quetelet and his acolytes, from Florence Nightingale, herself a significant statistician who thought social laws were imparted by God,[31] to Karl Marx, who used the work to support his materialist theories of change.[32] Even regular physicists were inspired by Quetelet's social physics. Renowned scientists like James Clerk Maxwell and Ludwig Boltzmann read and admired discussions of the iron laws of society, and used their statistical frameworks to model the movement of molecules. After all, these were only another kind of multitude, best understood through aggregate measures.

One particularly influential disciple of Quetelet was the English historian Henry Thomas Buckle, who expanded the scope of social physics to determine what he thought were the laws of history. In his 1857 work *The History of Civilization in England* (intended as the introduction to a multi-volume history of the world), he claimed that the fate of each nation was, to a large degree, determined by four basic elements: climate, food, soil, and what he called 'the General Aspects of Nature'. These attributes helped or hindered a country's development, but the last factor he considered predominant. It is a purposefully vague category that Buckle uses to refer to the aesthetic and practical challenges of the landscape: whether the terrain is comprised of inhospitable mountain and jungle or inviting hills and meadows. Buckle thought these factors gave 'particular tone to religion, arts, literature [and] all the principal manifestations of the human mind', and, from the point of view of his nationalistic project, handily explained the superior development of Europe. In many countries, he wrote, the 'pomp and splendour of Nature' has a tendency to overawe humanity, stimulating the imagination to excess, breeding superstition, and impeding rational thought. It was only in more restrained geographies like Great Britain's that the mind could assert itself, creating a confident race inclined to excel. 'Where the works of Nature are small and feeble, Man regains confidence; he seems more able to rely on his own power; he can, as it were, pass through and exercise authority in every direction.'[33] Thank God, in other words, for the plainness of England.

Buckle's work was hugely popular, going through multiple editions in languages including English, French, German, and Russian, and shows how statistical fatalism could interact with political attitudes. His work was shaped by the climate of laissez-faire liberalism that dominated the UK in the mid-nineteenth century, and he claimed that social regularities justify hard limits on government intervention – arguments that were repeated by leading politicians. It is the 'folly of lawgivers', wrote Buckle, to believe that change can be enacted in the face of ineluctable natural laws. 'No great political improvement, no great reform, either legislative or executive, has ever been originated in any country by its rulers.'[34]

Rebuttals to this wave of statistical fatalism took many forms: political, philosophical, and literary. Regularities hardly constituted laws, opponents argued, as while there may be consistent numbers of suicides or crimes each year, the exact cause of each event is unknown, meaning none are predetermined. Others suggested that although the physical world could be described in these terms, there was a Cartesian separation from the world of the mind that ensured human behaviour would 'defy every kind of calculation.'[35] One French critic compared statisticians' tables to skeletons without a body[36] – dead things that lacked the vital trappings of life. In the medical profession, where aggregated data was fast accumulating, doctors argued that statistics produced insights about the course and cure of diseases, but patients still had to be treated as individuals. As one Spanish doctor concluded: 'The law of the majority has no authority over refractory cases.'[37] As with the imposition of land surveys or standardised weights and measures, there is a tension here between the map and the territory, between the universal and the particular. In order to create a unified understanding of the world, it's helpful to ignore specifics. But such general rules are less persuasive when faced with individual suffering.

These arguments seeped into culture. In Dostoyevsky's *Notes from Underground* (1864), the narrator, a disillusioned civil servant, rails against experts' attempts to 'deduce the whole range of human satisfactions as averages from statistical figures and scientifico-economic formulas', while American essayist Ralph Waldo Emerson joked about the 'terrible tabulation of the French statists' that proved that if 'one man in twenty thousand, or in thirty thousand, eats shoes, or marries his grandmother, then, in every twenty thousand, or thirty thousand, is found one man who eats shoes, or marries his grandmother'. Perhaps the most enduring rebuttal of statistical fatalism, though, can be found in Charles Dickens's 1854 novel *Hard Times*.

Here, our protagonist Thomas Gradgrind, a school superintendent and later member of parliament, believes that 'facts alone are wanted in life. Plant nothing else, and root out everything else.' With his 'square coat, square legs, square shoulders', Gradgrind is a golem of statistical regularity: powerful, unyielding, and blind to the lives of those around him. When his son is caught robbing a bank, he rebukes his father for expecting anything better of him: '"I don't see why," grumbled the son. "So many people are employed in situations of trust; so many people, out of so many, will be dishonest. I have heard you talk, a hundred times, of its being a law. How can I help laws? You have comforted others with such things, father. Comfort yourself!"'[38]

Despite these criticisms, the development of statistical data undoubtedly spurred social reform. This was particularly true in the world of public health and insurance, where the ability to account for disease and accident on a mass scale led to greater interventions from governments and businesses. Often, the lessons of statistics were pleasingly direct in their effect. Consider, for example, the work of English industrialist Charles Booth, who surveyed the poor of London at the end of the nineteenth century, classifying every street, court, and block by the income of its residents, before rendering this data in colour-coded maps. Booth's work revealed that 33 per cent of the capital's population lived in poverty, rising to 35 per cent in the East End. These figures sparked public outcry and debate, and eventually led to legal changes, such as housing reform.[39] Here, the bird's-eye view provided by statistics was the only way to properly grasp the scale of poverty during the Industrial Revolution, convincing politicians to act when tales of individual misery had not swayed them. Coverage of Booth's work in *The Manchester Guardian* from 1889 praises him for lifting the 'curtain behind which East London had been hidden' and presenting the nation with a 'physical chart of sorrow, suffering and crime'.[40] When submitting his own work to the Royal Statistical Society, Booth said his aim was to create 'a true picture of the modern industrial organism' and so combat the 'sense of helplessness' felt by individuals.[41] Only by seeing the problem from afar, he said, could action be taken.

Correlation and regression

What the average meant to Adolphe Quetelet, the abnormal would mean to his most influential successor: the Victorian statistician Francis Galton. Galton recognised the importance of the normal curve, but found himself drawn to its margins: to the exceptional and the deficient. 'It is difficult to understand why statisticians commonly limit their inquiries to Averages, and do not revel in more comprehensive views,' he wrote in 1889. 'Their souls seem as dull to the charm of variety as that of the native of one of our flat English counties, whose retrospect of Switzerland was that, if its mountains could be thrown into its lakes, two nuisances would be got rid of at once.'[42] Born in 1822, a child prodigy from a prosperous Quaker family, Galton's career would span various scientific disciplines, from meteorology to geography and biology, and produce innovations as varied as forensic fingerprinting and the dog whistle. Today, though, he is most frequently remembered as the founder of eugenics, a doctrine that pushed the discipline of statistics to the most insidious ends imaginable.

Galton's work was founded on a single motive, a personal motto that he applied throughout his life with zeal: 'Whenever you can, count.'[43] One twentieth-century commentator notes that 'almost every scientific paper that Galton published was concerned, in some manner, with counting and measurement'.[44] Casual acts of quantification are littered throughout his life, including a measure of boredom in meetings based on fidgets per minute, the perfect temperature for brewing his tea,[45] and a 'beauty map' of Britain he created by 'classifying girls I passed in streets or elsewhere as attractive, indifferent, repellent'.[46] (Such objectification was familiar territory: during an expedition in his twenties to present-day Namibia, he recounts whiling away the hours using a sextant to measure the dimensions of Khoisan women from afar.[47]) He even tested the limits of his quantitative methods by applying them to matters of faith, publishing a study in 1872 titled 'Statistical Inquiries into the Efficacy of Prayer', in which he hoped to determine whether 'those who pray attain their objects more frequently than those who do not pray'. The evidence, he concluded, suggested otherwise. The most conclusive proof he offered was that despite an abundance of weekly prayers in churches for the health of the UK's monarchs, they remained 'literally the shortest lived of all who have the advantage of affluence'.[48]

However, it was the publication of *On the Origin of Species* in 1859 by Charles Darwin, a half-cousin of Galton's, that would provide the motivation for his life's work, stirring his interest in matters of heredity. Like many of his contemporaries, Galton was empowered by Darwin's conclusions about evolution by natural selection and felt that his theories offered a new axis upon which the world might turn. As Galton wrote to his half-cousin in 1869: 'I always think of you in the same way as converts from barbarism think of the teacher who first relieved them from the intolerable burden of their superstition.'[49]

In his writing, Darwin had only speculated on the mechanisms by which characteristics are passed from one generation to the next (the term 'genetics' wouldn't even be coined until 1905, years after his death), and so Galton turned to measurement and statistics to produce more definite knowledge. In his first substantial work on the subject, the 1869 book *Hereditary Genius: An Inquiry into Its Laws and*

Consequences, he proved to his own satisfaction that intellectual ability in humans was heritable. His source data was a biographical tome of 'eminent men' (Routledge's *Men of the Time*), which he rendered as family trees, categorising each individual into intellectual classes of his own devising based on the strength of their 'natural gifts'. He then analysed how the proportion of these classes, A through G, shifted in each family over time, concluding that 'each generation has enormous power over the natural gifts of those that follow'.[50] Following this line of thought, he mused that it would be 'quite practicable to produce a highly gifted race of men by judicious marriages during several consecutive generations', just as one might create 'a permanent breed of dogs or horses gifted with peculiar powers of running'.[51]

This was the crucial development that turned Galton's statistical observations into eugenics: his belief that not only could we measure humanity's virtues, but we could change them too. As he wrote in one article expounding the virtues of his new science: 'If a twentieth part of the cost and pains were spent in measures for the improvement of the human race that is spent on the improvement of the breed of horses and cattle, what a galaxy of genius might we not create! We might introduce prophets and high priests of civilization into the world, as surely as we can propagate idiots by mating crétins.'[52]

The incoherent aspects of Galton's reasoning in *Hereditary Genius* seem obvious now. His conclusions rest upon a number of flimsy assumptions that barely disguise his underlying prejudice. Why, for example, does he think that reputation is a reliable proxy for intellectual ability? Surely the offspring of the rich and powerful have material advantages over the children of the poor? Why should intelligence be heritable in humans just because horses can be bred to run slightly faster – are these qualities really equivalent? Reading

Galton now, it's easy to see how he mistakes the work of categorisation for more rigorous scientific enquiry. When he speculates on the intellectual differences between races, for example, he praises the ancient Greeks as 'a magnificent breed of human animals',[53] a few grades above modern Europeans in terms of their cognitive power. (It's safe, of course, to praise the dead in this way.) But 'the negro races', he says, are two grades below whites, a conclusion that Galton draws simply because he can think of no eminent Black people.[54]

This numerological approach is best captured by Galton's reverence for the normal distribution, which he saw, like Quetelet, as the stamp of nature's authority. If his data conformed to this pattern, it was 'a most trustworthy criterion' that his results were not shaped by 'dissimilar classes of condition' – that is to say, environmental factors[55] – but by inherent and heritable qualities. Galton would later coin the phrase 'nature versus nurture' to describe these two categories of influence, but was always clear which he thought dominant. As he wrote in *Hereditary Genius*: 'I have no patience with the hypothesis occasionally expressed, and often implied, especially in tales written to teach children to be good, that babies are born pretty much alike, and that the sole agencies in creating differences between boy and boy, and man and man, are steady application and moral effort.'[56]

———

Galton's veneration of the normal curve was such that he fantasised that it might one day take the place of the cross, if eugenics were ever to supplant mainstream religion as the fundamental organising principle of society. 'I know of scarcely anything so apt to impress the imagination as the wonderful form of cosmic order expressed by the "law of error",' he wrote. 'A savage, if he could understand it, would worship it as a god. It reigns with severity in complete self-effacement amidst the wildest confusion. The huger the mob and the greater the anarchy the more perfect is its sway.'[57]

But while Quetelet thought the average an admirable quality, Galton saw it as a menace. The hump of the normal curve was to him a looming wave of mediocrity that threatened to crash down on civilisation, drowning the few superior men and women capable of moving the world forward. Fear of this apocalypse animated many proponents of eugenics, but for Galton it also stimulated his mathematical curiosity. If the world depended on the success of the few, how could one separate the wheat from the chaff using the data alone? And if heredity played such an important part in distributing life's gifts, why was it so difficult to predict where the next genius would spring from? These questions would lead him to develop a great number of brilliant and extremely consequential statistical tools, for despite the blindness of much of his work, he was undoubtedly a mathematical prodigy.

After the publication of *Hereditary Genius*, Galton knew he would need more data to reveal the mechanics of inheritance. Prior to the invention of the IQ test there was no acceptable method for assessing intelligence, so he turned to biometric studies instead, creating his own methods to collect intergenerational data. ('The work of a statistician is that of the Israelites in Egypt,' he once complained. 'They must not only make bricks but find the materials.'[58]) He devised a battery of tests for human subjects, gauging everything from reaction time to punching strength to hearing. In 1884, he set up his 'Anthropometric Laboratory' – a gauntlet of health tests whose novelty attracted crowds of the public, who paid to be assessed and kept a copy of their data for the privilege. Galton was enough of a well-known figure at this point that the British prime minister at the time, William Gladstone, visited his lab, apparently convinced that he had an unusually large cranium. 'Mr Gladstone was amusingly insistent about the size of his head,' Galton noted later, 'but after all it was not so very large in circumference.'[59]

Years of collecting biometric data culminated in Galton's 1889 masterwork: *Natural Inheritance*. It was his most influential book, bringing together the insights of earlier years with wholly new statistical methods. In a study of intergenerational height, Galton noted that exceptionally tall or short parents don't necessarily have offspring with similarly extreme characteristics. Instead, their children's heights tended to be closer to the average. It was as if the slope of the normal curve held an inverse gravity, sucking in and subsuming outliers. Galton called this phenomenon 'regression towards mediocrity', though today it's known as regression to the mean. He then realised that the mathematical tools he'd used to track the difference between two expressions of height (that of parent and child) could be tweaked to measure the strength of connection between *any* two variables, to see if they moved in tandem or not at all. These measurements didn't have to belong to the same scale, but could encompass any set of figures, connecting anything from temperature and suicide rate to an individual's beard length and buoyancy. Galton named his method 'co-relation', though the spelling soon changed to the now familiar correlation.

The discovery of regression and correlation, writes Stigler, 'should rank with the greatest individual events in the history of science – at a level with William Harvey's discovery of the circulation of blood and with Isaac Newton's of the separation of light'.[60] They are tools of incredible flexibility and power, allowing scientists to draw conclusions from aggregate data that could never be found through other means. They amplify the power of relatively shallow measurements, allowing us to find connections between seemingly disparate phenomena. Take, for example, the connection between lung cancer and smoking, which is obvious today but poorly understood at the beginning of the twentieth century. Epidemiologists first spotted the correlation in population studies in the 1930s, but it took years for scientists to then establish a causal relationship based on carcinogenic chemicals in cigarette smoke. In complicated systems like human biology, where there are an incredible number of variants and interconnected systems, statistical tools are invaluable for steering investigation. As Galton's protégé and biographer Karl Pearson put it: 'Formerly the quantitative scientist could think only in terms of causation, now he can think also in terms of correlation. This has not only enormously widened the field to which quantitative and therefore mathematical methods can be applied, but it has at the same time modified our philosophy of science and even of life itself.'[61] And as Galton himself wrote: 'Some people hate the very name of statistics, but I find them full of beauty and interest. Whenever they are not brutalised, but delicately handled by the higher methods, and are warily interpreted, their power to deal with complicated phenomena is extraordinary.'[62]

Carrie Buck and IQ

It's curious to think what Carrie Buck might have made of this statement. Born in Charlottesville, Virginia, in 1906, five years before Galton died, Buck was definitely not 'delicately handled' by the great man's doctrine. Instead, her life shows what can happen when we begin to think of certain statistical measures as descriptors of innate human qualities, rather than features of the numerical landscape they are extracted from.

At the age of seventeen, Buck was raped by a nephew of her foster parents. She became pregnant and, as a punishment, was sent away to a state mental hospital. There, she and her newborn daughter were judged to be 'feeble-minded', a diagnosis that Carrie's own mother had received years before, and Buck was sterilised in the name of eugenics. We know from later testimony from her teachers that Buck was of normal intelligence, but the superintendent at her hospital wanted to make an example of her case.[63] His argument to sterilise Buck was pushed all the way to the Supreme Court, which in 1927 voted eight to one in favour of making forced sterilisation for eugenic purposes legal in the US. 'It is better for all the world,' wrote Justice Oliver Wendell Holmes Jr. in the court's majority decision, 'if instead of waiting to execute degenerate offspring for crime, or to let them starve for their imbecility, society can prevent those who are manifestly unfit from continuing their kind.' Holmes, often venerated as one of America's greatest legal minds, concluded with a brutal judgement on the Buck family: 'Three generations of imbeciles are enough.'[64]

What is perhaps most shocking about this case is quite how unexceptional it was at the time. The state of Virginia wasn't an outlier in its pursuit of genetic purity, but a laggard, with Buck's case only giving 'constitutional blessing' to a firmly established practice.[65] The first state laws on the matter had been passed two decades earlier, and by 1939 thirty-two states allowed such forced sterilisation.[66] A number of factors contributed to the rise of eugenics in America, including the country's established habits of racial hierarchy, social changes like mass urbanisation and immigration, and a reform movement that embraced eugenics as a scientifically sound form of societal improvement. As a result, the discipline was eulogised in popular writing, championed by the likes of Alexander Graham Bell and Theodore Roosevelt, and spread to the masses via grassroots movements. The American Eugenics Society, for example, made 'fitter family contests' a regular feature of state fairs in the 1920s, in which participants were judged on their supposed genetic virtues and handed medals stamped with the biblical motto 'Yea, I have a goodly heritage.'[67] It was just common sense that the future of the United States – indeed, of *all* Anglo-Saxon nations – relied on introducing laws that would 'dry up the springs that feed the torrent of defective and degenerate protoplasm.'[68]

The eugenics movement would not have existed without the statistical underpinning created by Galton and his disciples, but another form of measurement was equally crucial in justifying this creed: the IQ test. When Carrie Buck was judged by the state of Virginia as 'feeble-minded', the evidence was provided by the Binet-Simon test, the predecessor to all modern intelligence tests. It was invented in 1905 by French psychologist Alfred Binet and his colleague Théodore Simon after Binet was tasked by the French government to find a way to identify children who might need special assistance in the classroom. To achieve this, Binet created a series of tests to help identify a child's 'mental age'. Four-year-olds should be able to say which of two lines is longer and name household items, for example; seven-year-olds should be able to copy simple diagrams and describe pictures; ten-year-olds should recognise different coins and be able to make sentences from random words (like 'Paris', 'money', and 'river'). A few years later, a key feature was added: for ease of collating and comparing responses, test results would be rendered as a single number. This would be known as the subject's intelligence quotient, or IQ.[69]

IQ tests have since become perhaps the most influential and most misused social statistic of all time. Tragically, Binet himself never thought his test was a definitive measure of intelligence, or even that it should be used to rank cognitive ability at all. He conceived of it simply as a tool to identify who needed help in the classroom. It was only later practitioners who, like Quetelet and his averages, turned it into a 'real' quality possessed by individuals. Binet himself was adamant that cognitive ability was too complex to be measured by a single unit, and even foresaw how his work might be misused. 'We feel it necessary to insist on this fact,' he wrote in 1911, 'because later, for the sake of simplicity of statement, we will speak of a child of 8 years having the intelligence of a child of 7 or 9 years; these expressions, if accepted arbitrarily, may give place to illusions.'[70] Indeed, Binet predicted many of the problems associated with IQ tests: that they can be used as an excuse to ignore children who are difficult to teach, for example, and that the results are seen as a permanent assessment of a child's ability rather than a snapshot in time. Binet stressed time and time again that his tests assessed only 'the capacity to learn and to assimilate instruction',[71] rather than any innate or fixed quality.

This caution was soon abandoned when Binet's tests arrived in the United States. The psychologist and eugenicist Henry Goddard was the first to adapt his work, translating the tests into English in 1908. Goddard presented these examinations as the definitive measure of an individual's intelligence, writing: 'Each human being has a potentiality for a definite amount of intelligence [...] beyond that point all efforts at education are useless.'[72] In 1910, he proposed a new tripartite system for classifying individuals with low IQ. 'Idiots' were those with a mental age of less than a three-year-old; 'imbeciles' had a mental age of between three and seven; and 'morons' (an original coinage of Goddard's) had the intelligence of someone aged between eight and twelve. Crucially, he stressed that these categories were immutable. Goddard did not, like Binet, consider that one's ability to take tests might reflect numerous factors, like access to education or trust in authority, but declared that intelligence simply *was*. '[N]o amount of education or good environment can change a feeble-minded individual into a normal one, any more than it can change a red-haired stock into a black-haired stock,' he said.[73]

Such ideas became extremely influential in the United States, and Goddard's work was used to justify xenophobic and racist assumptions. In 1913, he began testing immigrants arriving at Ellis Island, focusing on those travelling in steerage, the cheapest accommodation. He found that 79 per cent of Italians, 80 per cent of Hungarians, 83 per cent of Jews, and 87 per cent of Russians were 'feeble-minded', results so shocking that he adjusted the test's parameters to be more generous. Even then, he concluded that 40 to 50 per cent of immigrants had a mental age less than that of a twelve-year-old. 'We cannot escape the general conclusion', said Goddard, 'that these immigrants were of surprisingly low intelligence.'[74] The fact that the people he and his assistants interviewed were arriving in a new country after weeks of travel crammed like cattle into a ship's hold,[75] and probably had a less than perfect grasp of the language, did not factor into his assertions.

Such dismal results, though, were not anomalous. In 1917, an IQ test also adapted from Binet was administered to roughly 1.75 million US Army recruits. It found that the average mental age of a white recruit was 13 years old; for a Russian, 11.34 years; an Italian, 11.01; and for a 'Negro', 10.41 years (which test coordinators further subdivided by lightness of skin colour, claiming that Black people with lighter skin were more intelligent). Again, environmental factors like the poor education of many recruits and the culturally dependent nature of the test's questions were not considered, and the apparent low intelligence of the average white recruit became 'a rallying point for eugenicists, who predicted doom ... caused by the unconstrained breeding of the poor and feeble-minded'.[76] The accumulation of results like these were shocking enough that even Goddard, who would later recant much of his early research, began to question his methods. 'We seem to be impaled on the horns of a dilemma: either half the population is feeble-minded; or 12 year mentality does not properly come within the limits of feeble-mindedness.'[77]

'A few modern philosophers [...] assert that an individual's intelligence is a fixed quantity, a quantity which cannot be increased. We must protest and react against this brute pessimism.'[80]

In the case of eugenics, the doctrine only fell out of favour after the end of the Second World War, when the true extent of the Nazis' barbarous sterilisation programme was revealed. More than 60,000 people were forcibly sterilised in the US, while some 375,000 were sterilised in Germany, with millions more murdered in an effort to 'cleanse' Europe of those deemed to be *Lebensunwertes Leben* – an official category that translates as 'life unworthy of life'. These aren't issues of the past, either; reports in 2020 revealed that hundreds of women were forcibly sterilised in Californian jails right up until 2014,[81] while the repressed Uyghur minority in China has been similarly subjected to involuntary sterilisation, birth control, and abortion.[82] At the Nuremberg trials, when members of the Nazi Party were prosecuted for their crimes by the Allies, Otto Hofmann, head of the SS Race and Settlement Office, defended his actions by pointing out that eugenics was hardly a German creation.[83] Indeed, the country which the Nazis had looked to for inspiration in these matters was the United States itself. Hofmann quoted the words of Wendell Holmes Jr. in *Carrie vs Buck* back to his prosecutors: 'Three generations of imbeciles are enough.' The Nazis, he said, only followed these teachings to their logical end.

8

THE BATTLE OF THE STANDARDS

———————

Metric vs imperial and metrology's culture war

Then down with every 'metric' scheme
Taught by the foreign school.
We'll worship still our Father's God!
And keep our Father's 'rule'!
A perfect inch, a perfect pint.
The Anglo's honest pound.
Shall hold their place upon the earth.
Till Time's last trump shall sound!
—SONG OF THE INTERNATIONAL INSTITUTE
FOR PRESERVING AND PERFECTING WEIGHTS
AND MEASURES[1]

Active Resistance to Metrication (ARM)

Casing the joint is an essential part of any heist movie. The protagonists arrive on the scene, dressed suitably incognito, and do their best to blend in while keeping one eye on their target: an historic painting with emotional significance for the hero, perhaps, or a jewel of unrivalled clarity and brilliance, too tempting not to steal. These were the sorts of treasure floating through my mind as I sat in a pub on a dull October afternoon, reconnoitring a much more mundane target: a wrought iron signpost.

The signpost, standing on the other side of the street and quite unaware it was under careful surveillance, couldn't lay claim to much historic significance or monetary value. In fact, it wasn't of much interest to anyone unless they happened to be passing through the town of Thaxted in south-east England and wanted to know how many metres it was to the historic windmill. What made this signpost noteworthy, though, was that it had come to the attention of a metrological vigilante group known as Active Resistance to Metrication, or ARM. Their sworn goal? To 'oppose forced metrication' in the UK and preserve the country's traditional imperial units. Their strategy? To wage a guerrilla war against metric road signs and signposts: unscrewing them in the dead of night, stowing them in hedgerows, or amending them using paints and stickers. Gone were the metric monstrosities and in their place were the wonderful, practical, and *British* units of miles, yards, and feet.

Since ARM was formed in 2001, its members have removed or altered more than 3,000 signposts across the country, striking in country villages, seaside towns, and even the nation's capital. The group, citing the Department for Transport's Traffic Signs Regulations, says its work is perfectly legal, but its operations are

For the UK's Active Resistance to Metrication organisation, retaining traditional units of measurement is a battle for the nation's soul

nevertheless performed in a clandestine manner. Members adopt *noms de guerre* for their work, choosing names inspired by obsolete units, like Rod Perch and Polly Peck, and deploy specially made gadgets to complete alterations discreetly (one favourite tool is a stepladder in a briefcase known as 'Henry the Height Enhancer'). Despite this silliness, their targets don't see the funny side. Members of ARM have been arrested and prosecuted for their actions, even spending the odd night in jail.

This was the legend of the group I'd read about online; but sitting opposite me in the pub on that October afternoon, the aforementioned signpost visible over his shoulder, was, I was beginning to suspect, the totality of ARM's active membership. His name was Tony Bennett (codename: Hundredweight) and he was sipping a pint of cider while explaining to me the link between his evangelical Christianity, Euro-scepticism, and this idiosyncratic campaign against metric measures.

When Tony first met me off the train from London, he was sitting on a low wall and eating an early lunch out of a faded plastic ice-cream container. He struck me as a kindly figure, his old-fashioned glasses, V-neck jumper, and stiff gait reminding me of my own grandfather. The more we talked, though, during strolls around Thaxted's idyllic cobbled back streets and over pints in its various pubs, the more his own character pushed past my preconceptions. Tony was a fanatic, I realised, albeit a quiet and bookish one equipped with clipboard and notebook.

It all went back to Nimrod, he was saying. Nimrod, great-grandson of Noah and the 'mighty hunger before the Lord', had attempted to unite the world's population, building the Tower of Babel so that humanity might climb up to Heaven itself. 'And God intervened, stopping him from building the tower,' said Tony. 'And in order that people should spread across the globe and reap the benefits of the Earth's variety, he gave them different languages – in an instant, supernaturally.' This, he says, was how the world was divided into different nations, which is God's preferred set-up. 'By and large, we Christians, we do believe that the nation state is a stable force for good,' says Tony. 'People should live in distinct nations because it provides a unifying force in their lives. It gives them a sense of purpose.'

That purpose led Tony to join the UK Independence Party in 1997 – a first step towards his interest in traditional weights and measures. 'I saw them campaigning in Harlow and thought to myself, "Weirdos," but I picked up one of their leaflets anyway,' he says. He would later become a solicitor for the party and political secretary to Jeffrey Titford, its leader from 2000 to 2002 and one of the first three UKIP Members of the European Parliament. The literature he received from UKIP and its intellectual base contained revelations to match those of scripture, showing him 'that in actual fact the European Union project was not by any means a trading union, but a well-laid plan going back to the 1940s to remove the powers of the nation state'.

'When I came to look closely at it,' he told me later, 'the more it appeared to me that the European project was a deliberate attempt to reverse what happened at Babel. To say that the idea of the nation state is redundant, and that what we need to build is a strong international organisation, perhaps even a one-world government.' Fighting against this tendency meant not only getting out of the EU, he says, but combating other aspects of pan-European integration, from the adoption of a single currency, the euro, to the use of metric weights and measures across the continent. For Tony and his compatriots, the rallying cry of the French revolutionaries in the *Cahiers de doléances* for 'one king, one law, one weight, and one measure' had returned as a threat.

For ARM and its more legitimate cousin, the British Weights & Measures Association, or BWMA, the battle against metrication in the UK began in 1965, when the government's Board of Trade outlined a ten-year scheme to go metric. The plan was drawn up at the behest of industry groups, who felt that the UK's attachment to imperial units was holding it back in the international market. Importantly, this decision to metricate was taken years before the UK joined the European groups that would form the nucleus of the European Union, showing that metrication in the UK has been, from the start, an internal decision rather than an external imposition. In the years that followed, various aspects of everyday life moved over to metric. Paper sizes in 1967, pharmacy prescriptions and higher education exams in 1969, and so on. In 1971, the country took the significant step of metricating its currency, dropping the old system of pounds, shillings, and pence (itself inherited from the work of the eighth-century metrological reformer Charlemagne) in favour of decimal divisions. This changeover, as culturally significant as the loss of pounds and ounces, happened without too much trouble, and in the decades that followed, all the UK's major industries, from car-making to pharmaceuticals, went metric. By the 1990s, only a handful of prominent public-facing measures, like road signs and grocery weights, remained in imperial or dual units.

The slow march of the metric system might have continued unabated if it hadn't been for a bunch of bananas. In the year 2000, a market-stall owner in Sunderland named Steve Thoburn sold the fruit in question to an undercover trading standards officer, pricing them using pounds and ounces (25p for a pound) and so contravening an EU directive that all loose goods should be sold using metric units. Thoburn's scales were confiscated and, along with four others accused of similar crimes, he was convicted in 2001 of breaking the law.[2] The case raised the question of weights and measures to national prominence and the group was dubbed the 'Metric Martyrs' by the press, after a spokesperson for the Institute of Trading Standards Administration had said the men could martyr themselves if they wanted, but it wouldn't change the law.[3] The headlines were irresistible – 'Market man faces scales of justice' – and when the traders appealed their conviction at the UK's High Court, UKIP saw an opportunity and helped cover Thoburn's legal fees, mobilising its supporters and staging protests outside the hearing. They waved banners with slogans like 'Our Weigh Is Better' and 'Rule Britannia – In Inches Not Metres'. At one protest, they even set up an impromptu fruit and veg stand outside the court.[4] Bunches of bananas were sold to fellow protestors, all weighed out in good old-fashioned imperial, of course.

As Tony explains, it was the perfect wedge issue with voters, one that UKIP knew would attract people to their cause. The case was simple, easy to relate to, and succinctly captured fears about EU overreach. 'It became part of popular folklore,' he says. 'You'd have right-wing commentators saying, "How can it be that Steve Thoburn can't sell a pound of bananas?" People rang up and joined the party after reading about it, saying, "Finally, somebody's standing up to these Eurocrats"' (even though, of course, it was the UK government's implementation of EU law that had led to Thoburn's prosecution, not the EU directive itself). And although the case seems frivolous, it helped set a legal precedent, establishing the supremacy of EU law over UK legislation in certain areas. The judge overseeing the appeal even described the disputed fruit as 'the most famous bunch of bananas in English legal history'.[5]

Among those who joined the media scrum was future UKIP leader and Brexit champion Nigel Farage, who had been elected as one of the party's first three MEPs in 1999. When the traders lost their appeal in 2002, Farage fulminated on the radio and in the papers, presenting the case as the end of the UK's political autonomy. 'What more proof is needed that the UK is now ruled by the EU, and that Parliament has been rendered useless?' he asked.[6]

For years, such complaints were a relatively fringe matter, but the discontent they captured was more widespread than many realised. When the UK voted in 2016 to leave the European Union, one BBC political reporter identified this dispute over weights and measures as a watershed moment for the Brexit project: an event that 'helped turn public opinion against EU membership, giving critics something tangible to point to that affected people's everyday lives and for which Brussels appeared responsible.'[7] As with the *pietre di paragone* carved into the marketplaces of Italian towns, or the meticulous rules about measuring grain in medieval Europe, arguments about the price of bananas aren't abstract or academic – their significance is weighed in front of your eyes. And the grievance had remained, even though the European Union relented on the issue. In 2007, in fact, the EU had told the UK it could keep using imperial measures wherever it liked. As Günter Verheugen, EU industry commissioner, said at the time: 'I want to bring to an end a bitter, bitter battle that has lasted for decades and which in my view is completely pointless.'[8] Today the UK is almost entirely metric, but retains dual units on some food packaging and imperial measures in areas of life too culturally embedded to suffer change. There are still miles, yards, and feet on road signs; most people measure their height in feet and inches, and many use stones, pounds, and ounces to track their weight. And no one would contemplate getting rid of pints in pubs. It's in these measures – close to the heart and the body – that imperial units are securely moored.

As Tony explains all this to me over our own pints in Thaxted, I eye the signpost outside that we are supposed to 'amend'. It's obviously a stupid thing to do, I think, changing units of measurement on a signpost. A pointless act of minor vandalism that will only annoy the council worker who has to fix it. But in that case, why did Tony care? Why had he been doing this for so many years? It couldn't just be about the attention (though it was definitely about that too).

We chatted some more as locals from the village began filing in, each taking a single seat at a distant table in line with the latest COVID-19 restrictions. Their noise punctured the intimate atmosphere that had settled between Tony and me, as they shouted at one another with relish across the room and called for drinks from the bar staff. Whenever one left their seat and veered towards someone else's table, they were met with a hail of cheerful outrage. 'Two metres apart, I said *two metres apart*, Steve, don't you dare come any closer to me!' This was official government guidance: metres not feet. As I pointed this out to Tony, curious if he'd noticed or even cared, he merely grimaced at me over the remains of his pint. 'We should drink up,' he says. 'We need to get this done before the light goes completely.'

Uniform, stable, unworkable

The history of anti-metric sentiment is as old as the metric system itself. Indeed, you could argue it began before metric weights and measures were even finalised, as French intellectuals debated whether or not the proposed names for their new units – like the *millimetre* and *decicadil* – would be a hurdle to public adoption. In the UK and US, though, metrication has been met with particular suspicion and fear. This is in part due to the usual challenges of implementation, but also because these countries created and inherited the only system of measurement to offer a viable challenge to metric: British Imperial and US Customary measures. (There are small differences between these systems, particularly in how they measure volume, but in history and usage they are more alike than not, and for the sake of simplicity I will talk about imperial measures going forward.)

In the US, metric overtures were present from the country's founding. The scientifically minded Thomas Jefferson had long been agitating for reform in weights and measures and closely followed the work of the French *savants*, while George Washington himself noted in his inaugural address to Congress in 1789 that 'Uniformity in the currency, weights and measures of the United States is an object of great importance, and will, I am persuaded, be duly attended to.'[9] But this approach to metrological reform – stressing its great importance while suggesting that someone, *anyone*, but the speaker should actually address the problem – would become a familiar motif on both sides of the Atlantic. The necessity and inevitability of metrication is recognised time and time again in these nations (as it is to this day), but few seem willing to inflict the short-term pain necessary to reap the long-term rewards. Three years after Washington's address, money in the US achieved 'uniformity' with the Coinage Act of 1792, which established the silver dollar and its decimal divisions, but the issue of weights and measures would take much longer to sort out.

It was Jefferson who was first tasked with investigating how the US might reform its weights and measures on his return to the country as Washington's Secretary of State in 1789. As with the implementation of his grid survey, Jefferson hoped that a unified system of measurement might create a more unified nation. Although most trade on the continent was conducted in units inherited from the English, colonists from different nations had brought their own measures to the US, creating a hotchpotch system of variants from Ireland, Scotland, the Netherlands, and elsewhere[10] that hampered trade and commerce, just as it had in Ancien Régime France.

Jefferson's own interest in the subject went beyond the political, and the man seems to have had a keen personal interest in the quantified life. One of his fancies was buying custom-made odometers for his carriages that measured how far the vehicle travelled as he criss-crossed the country on government errands. The odometers fitted into the spokes of the carriage wheel and counted each rotation, letting Jefferson calculate the distance travelled from measurements of the wheel. He obsessively tallied these distances, recording journeys in tables that begin 'from Monticello' and list dozens of stops before arriving at the 'president's house.'[11] As a man of science, he was proud that his calculations were decimal, writing in a draft of his autobiography: 'I use, when I travel, an Odometer of Clarke's invention which divides the mile into cents [hundredths], and I find every one comprehend[s] a distance readily when stated to them in miles & cents; so they would in feet and cents, pounds & cents, &c.'[12] Later in life he upgraded to a model that automatically rang a bell with each passing mile, writing later to its inventor that he knew 'no Odometer either in Europe or America [...] to be compared with it' and that this particular feature gave him 'great satisfaction.'[13]

Jefferson's investigations into the metric system did not run quite as smoothly, though. The first and most unexpected of his problems was British pirates, or, to follow the niceties of nineteenth-century state-sanctioned plunder, privateers: private ships that raided foreign waters and handed over a slice of their profits to the crown. When Jefferson requested a copy of the standard metre and kilogram from France in 1793 (the kilogram then being called the *grave*), the ship carrying the artefacts was blown off course during its voyage across the Atlantic, drifting into the Caribbean Sea, where it was seized by raiders. The significance of the standards was lost on the pirates, who auctioned them off along with the rest of the ship's contents.[14]

The standards, which might have helped make the case for metric conversion, never made it to Jefferson's care, but this loss wasn't necessarily the deciding factor. Instead, when Jefferson investigated the scientific basis of the metric system, he found it unconvincing. Using the Paris meridian to define the unit of length worried him, with recent experiments showing that the Earth wasn't a perfect sphere but was squashed at the top and bottom like a tangerine. This meant not every segment of the meridian would follow the same curve, and so the Paris meridian, and *only* the Paris meridian, could be used to capture and recreate the metre's value. This localisation, wrote Jefferson, 'excludes, ipso facto, every nation on earth from a communion of measurement with [the French].'[15] As a result he declared the metric system 'unworkable'. Based on this analysis, Jefferson offered Congress two proposals in 1790: they could either redefine English units using a system of Jefferson's devising (though based on the established science of using the swing of a pendulum rod to create a length standard) or simply 'define and render uniform and stable' the country's weights and measures by distributing new standards to each state.[16]

In the end, the government chose to do nothing. Jefferson's proposals would bounce around the Senate and Congress for years, their passage trailed by the formation of well-intentioned committees, but no politician seemed willing to take the plunge and decide on a course of action. When John Quincy Adams had the problem dumped on his desk as Secretary of State in 1817, he found it no more amenable. He concluded that reforming weights and measures was 'one of the most arduous exercises of legislative authority', not because of the difficulties of enacting the law, but of 'carrying it into execution'. To change all the units of a country at a single stroke was to 'affect the well-being of every man, woman, and child, in the community. It enters every house, it cripples every hand.'[17]

This pervasive nature of measurements helps to explain why changes to units so often occur in times of social upheaval, such as conquest or revolution. It is only during these moments, when old sureties are tossed into the air like dice to fall who knows how, that reordering anything as fundamental as measurement can take place. In France, for example, although Napoleon had thrown out the metric system in 1812 in favour of his hybrid *mesures usuelles*, the kilogram and metre returned to the country with the July Revolution of 1830. This saw the conservative King Charles X replaced by his more liberal cousin, Louis Philippe I, who justified his rule by promising to restore the progressive sensibilities that birthed the Revolution without the attendant bloodshed. Reinstating the metric system was a symbol of this salvage: the rescue of a great intellectual achievement from the gore and debris of radicalism. It didn't hurt either that Philippe's court was full of bankers, landowners, and industrialists, all of whom would benefit from the harmonised measures of the metric system.

Elsewhere in Europe, the metre and kilogram took root as a result of Napoleonic conquests. Metric measures 'marched in the wake of French bayonets', as Witold Kula puts it, and were imposed on countries along with the various legal and commercial reforms of the Napoleonic Code.[18] Some nations objected to the new measures, but others welcomed them. In the Netherlands, Belgium, and Luxembourg, for example, metrication was seen as a useful replacement for the tangle of native units, but in the Italian peninsula, where city states had strong histories of commercial independence and local measures, resistance was much stiffer.[19]

Two powerful political concepts, though, would help entrench metrication more widely in the nineteenth century: the twin creeds of nationalism and internationalism. For a certain sort of nationalist, metrication appealed as another transcendent offshoot of Enlightenment thinking: a universal system of measurement to accompany the universal rights of man enshrined in new political constitutions. For the more pragmatically minded, metric weights and measures offered the means to erase regional disparities, binding together disparate economies and industries into a single national project. For these reasons, metrication was favoured in otherwise contrasting forms of nationalist thought, adopted during both the liberal *Risorgimento* in Italy, where metric units were introduced with the formation of the Kingdom of Italy in 1861,[20] and the more authoritarian process of unification of Germany, where states like Prussia and Bavaria began using metric-based units to create a common language for trade before the creation of the German Empire in 1871.[21] In South America, too, metrication was seen as part of the project of nation-building: a neutral and rational system that was adopted by a string of countries in the 1860s and 1870s after decolonisation. These included Chile, Mexico, Brazil, Peru, Colombia, Uruguay, and Argentina. As the Italian statesman Massimo d'Azeglio commented in 1860, prior to the country's unification: 'We have made Italy. Now we must make Italians.'[22] Metrication was one of many tools it was hoped would achieve this.

At the same time that nationalists were finding reasons to love metric units, the system was also being proselytised by international thinkers. Internationalism had sprung from nationalism, uniting intellectuals and politicians of many different stripes. Socialists, liberals, and imperialists alike aspired to knit the world closer together, whether through solidarity of the workers, lower barriers to trade, or the conquest of new territories. And all saw the potential of new inventions – of steamships, railways, electricity, and telegraphs – to achieve that aim. The ebullient spirit of the time is exemplified by the International Peace Congress: a series of meetings organised by activists from France, Germany, the US, and the UK that created a template for future international peace efforts.[23] At the 1849 Congress in Paris, French novelist and poet Victor Hugo whipped a crowd into a hat-waving frenzy with what his biographer describes as a wide-eyed paean to technological advances and the glorious possibilities of the human race. 'See what discoveries are every day issuing from human genius – discoveries which all tend to the same object – Peace!' roared Hugo. 'What immense progress! What simplification! How Nature is allowing herself to be more and more subjugated by man!'[24]

These beliefs helped make the case for an international system of weights and measures, with many groups throwing their support behind the metric system during this period. These were often professional organisations, coalitions of statisticians, geographers, and the like whose work would benefit from the simplicity of metrication. In 1863, for example, at the Paris Postal Conference, a group of fifteen countries, responsible for sending and receiving 95 per cent of the world's correspondence, agreed to use metric measures for postal weight classes. As one delegate told his peers at the meeting, the mail they oversaw would 'diffuse the printed elements of civilization, progress, and intelligence. In each of these ways, they serve to break down the useless barriers which ignorance and non-intercourse formerly interposed between nations.'[25] Weighing by the kilogram would only help this good work.

The metric fever was so great that even the UK and the United States came close to conversion during this period. In Britain, a bill for metric adoption passed through the House of Commons by 110 votes to 75 but never made it to the House of Lords because of time constraints, and was rejected in a second vote by a majority of five in 1871.[26] Politicians' enthusiasm for metric surprised many, with an editorial in The Times from 9 July 1863 expressing outrage and disbelief in the plans. It declared that conversion would fill every household in the country with 'perplexity, confusion, and shame'.

In America, the Metric Act was passed in 1866, legally protecting the use of metric weights and measures in industry and providing official conversion tables for local units (a similar law was passed in the UK in 1864). The bill's sponsor, Republican abolitionist Charles Sumner, championed it on internationalist principles, declaring in a speech to the Senate that there was 'something captivating in the idea of one system of weights and measures, which shall be common to all the civilized world, so that at least in this particular, the confusion of Babel may be overcome. Kindred to this is the idea of one system of money. And both of these ideas are, perhaps, the forerunners of that grander idea of one language for all the civilized world.'[27]

Then came the most significant step towards the metric system's global pre-eminence: the 1875 Treaty of the Metre, which created a number of organisations charged with defining, developing, and promulgating metric measures. These include the International Bureau of Weights and Measures (known as the BIPM after its French title, the Bureau International des Poids et Mesures), which coordinates metrological research across different nations. The BIPM also centralised the creation and distribution of metric standards, issuing each of its original seventeen signatories with freshly forged kilogram and metre standards. Notably, the US was one of these seventeen signatories, but the UK was not. As one historian puts it, the metric cause in this period was regarded by many elites as 'so worthy and ameliorative as to be above denial, almost beyond criticism.'[28]

Pyramid inches and 'our inheritance in the desert'

When the elites of the world gather together, seemingly unified around a single cause, it's bound to create suspicion. And as fervour for metrological reform returned to US shores in the second half of the nineteenth century, it was met by an upsell of anti-metric sentiment. As historian of science Simon Schaffer has noted, metrology in this period was shouldering a particularly heavy burden, with measurement expected to 'resolve simultaneous polemical demands for pious morality, capitalist economy, and scientific accountability'.[29] As a result, debates about the origin and maintenance of measurement standards were particularly heated, as various groups projected their demands and concerns on to these units.

In America, anti-metric sentiment found an outlet in the International Institute for Preserving and Perfecting Weights and Measures, a small group that was the first official anti-metric organisation, founded in 1879.[30] The group encompassed a range of characteristics and beliefs, including a distrust of foreigners, class anxiety, and a love of pseudoscience, but its hundreds of members rallied around a single cause: the defence and preservation of traditional measures. During its brief but intense nine-year lifespan, the Institute wrote poems and pamphlets, gave speeches and lobbied politicians, and even had its own theme song, a rousing number titled 'A Pint's a Pound the World Around' that combined jingoism and faith in stentorious iambic measure, calling for an end 'to every "metric" scheme' and triumph for 'the Anglo's honest pound'.

The song also references one of the most popular metrological myths of all time: that the inch, pint, and pound are, in some way, divine units of measurement bestowed by God. This belief is still popular in fringe circles today but can be traced back to one of the nineteenth century's greatest fads: the pseudoscience of pyramidology. This term now refers to a number of conspiracy theories centred on the pyramids of Egypt, covering everything from alien architects to mysterious pyramid power, but in its earliest form it was firmly metrological. It was the belief that the Great Pyramid of Giza is built to God's specifications and that, if one properly measures its dimensions, the 'silent testament' of the stones will reveal not only the history of the world, but its future too.

The father of pyramidology was one John Taylor, a London publisher, lifelong bachelor, and steward to a smattering of poetic greats, including John Keats. In 1859, Taylor published the first book to claim at length that the pyramid's design was the product of divine instruction, titled *The Great Pyramid: Why Was It Built? And Who Built It?* Taylor found that if you divide twice the length of the structure's base by its height, you get a figure that is exactly *pi* – an irrational number and mathematical constant not formally discovered until centuries after the pyramid's construction. Taylor suggested that the pyramid had been built using a 'sacred cubit' as a base measure, tracing this theory back to Isaac Newton himself, who had claimed the same unit was used in the construction of Noah's Ark, Solomon's Temple, and the Holy of Holies, where the Ark of the Covenant was kept.[31] Divide this 'sacred cubit' by 25 and you get the British inch (more or less), leading Taylor to proclaim that the pyramid had been built by Israelites guided by the 'Great Architect' himself, who intended for its stones to preserve this sacred measure for use by his chosen people (the Anglos).

Taylor's theories were vivid and inventive, but would have remained largely unknown if not for one man: Charles Piazzi Smyth, a respected and accomplished scientist who had been appointed as Scotland's Royal Astronomer at the age of just twenty-six. Smyth would become the greatest proselytiser of Taylor's vision, expounding on his theories in a series of books, beginning in 1864 with the 664-page *Our Inheritance in the Great Pyramid*. Unlike Taylor, Smyth made a trip out to Egypt to examine the Great Pyramid in person and take his own measurements. He travelled with his wife and lived in an abandoned tomb on the Gizan plain, publishing his findings in a three-volume work titled *Life and Work at the Great Pyramid*, where he concluded that in the same way that God had gifted Adam with language, the Great Pyramid was 'devised likewise for the metrology of all nations; and has been by them unconsciously so employed to a great extent'.[32]

Collectively, these books were a sensation. They were read by millions, translated into many languages, and provided the leaping-off point for all sorts of pyramid-focused theories. With Taylor's sacred cubit in hand, Smyth surveyed every aspect of the great tomb, finding an abundance of measurements he thought corresponded with scientific and historical facts. Everything from the mean density of the Earth to the length of the polar axis and the distance from our planet to the sun was written into these stones. He declared that the empty granite sarcophagus in the King's Chamber was actually a measurement of capacity, and developed a theory that the pyramid's main internal corridor, the Grand Gallery, constituted a hidden history of the world. If you measured the passageway in sacred inches, with each inch corresponding to a year, it became a calendar of creation, with certain signs – like change in incline or a mark on the wall – indicating historical events like the Deluge and the birth of Christ. Somewhat worryingly for Smyth, the Grand Gallery was not particularly long, and if it really did constitute a history of the world, would it not also show when the Apocalypse would take place? 'The answer to that question', he wrote, 'must depend solely on the length of the Grand Gallery by measure in Pyramid inches.' His initial calculations marked a date of 1881 for the end of the world, but this was a little close for comfort, and in a later publication he suggested that the passages and chambers at the end of the gallery might also be considered part of this sacred calendar. He interpreted a particularly narrow passage following the gallery, which had to be passed through bent over double, as the period of strife said in the Bible to precede the Second Coming of Christ. 'But this excruciating portion is exceedingly short,' he comforted readers, and after fifty-three years of struggle humanity would enter 'into the freedom of the Ante-chamber, its quiet, peace and presently *granite* protection'.[33]

These calculations proved to Smyth that the Great Pyramid was the product of something 'much more than, or rather something quite different from, human intelligence'.[34] But he also made clear that his allegiance to imperial units was founded on other beliefs, including his religious convictions and suspicion of foreigners. '[W]ith the elevation of the metrical system in Paris,' he wrote, 'the French nation (as represented there) did for themselves formally abolish Christianity, burn the Bible, declare God to be a non-existence, a mere invention of the priests, and institute a worship of humanity, or of themselves'.[35] He also repeatedly stressed the superiority of European culture and science, and its ability to uncover hidden truths ignored by the Egyptians themselves. He complains that, to the natives, his clinometers and theodolites were 'proofs in their minds that a European cannot get on at any occupation without some queer and troublesome contrivance to *peep through*'.[36] They were not exactly wrong.

———

This talk of ancient wisdom and divine purpose appealed greatly to the pious and self-aggrandising metrologists of America's International Institute. One of the most devout members of the group, Charles A. L. Totten, a professor of military tactics and composer of the organisation's pints-and-redemption theme song, pushed Smyth's rhetoric even further. In his 1884 book *An Important Question in Metrology, Based upon Recent and Original Discoveries*, he declares the Great Pyramid a 'cabalistic symbol of the earth itself saved from the greater flood' and urges readers to listen to 'the momentous truths with which its metrological proportioned blocks reply, in cosmic ratios, to the grand dimensions of the earth on which they live'.[37] He also amplifies the xenophobic strains in Smyth's arguments, praising the United States as a nation 'marked out by special manifestations of Divine Providence' and uniquely deserving of that 'inheritance in the desert'.[38] American measures are simply what American men and women deserve, Totten trumpets: 'Our representatives have no more right to force the metric system upon us than they have to make our babies beg for bread in foreign idioms.'[39]

It can be hard to imagine how units of measurement can sustain these sorts of passions, but Totten's book is revealing in this regard. A large part of it is filled with dense tables of units, showing how ancient weights and measures relate to one another and their contemporary equivalents. It feels like a ritualistic chant or mantra, with Totten fusing the mysticism of numerology with the revelations of conspiracy. At one point he calculates a series of liquid measurements,

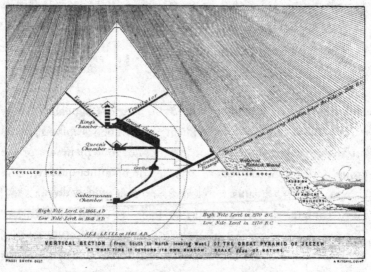

An illustration from Smyth's 1864 book *Our Inheritance in the Great Pyramid,* which dissects the pyramid for geometric wisdom

starting with the smallest unit, a drop, and multiplying it through progressively larger and unusual units – logs, baths, lavers, and so on – until he reaches his final measure of liquid volume: '1 sea'.[40] It's an approach that's similar to the Onomasticon of Amenopĕ, putting the furthest reaches of the world into simple order, and shows the intoxicating potential of measurement: its ability to make chaos coherent and to encompass vast spans of the world.

Such a combination of science and mysticism is enthralling, but it's not what won the nineteenth century's Battle of the Standards, as the rivalry between metric and imperial became known. Despite their love of pseudoscience, Smyth and the International Institute also outlined more sensible reasons as to why the US did not need the metric system, which made their way into both public and political discussions and still appear in debates today.

The most common of these arguments is the perceived appropriateness of imperial units. They are measures that have developed slowly over time and have become shaped, like a familiar tool, to the user's grip. Practically speaking, the biggest difference is that these measures are not decimalised. In decimal systems one relies on multiples of 10, making it easy to convert between units and calculate with large numbers. But US customary measures and British imperial measures use either 12 or 16 as their base (for example, 12 inches in a foot and 16 ounces in a pound), which is easier for dividing things into thirds and quarters. It's this feature that makes these systems more appropriate to market bartering and other quick calculations. Think of it like slicing up a pizza. You can quickly divide it into quarters or eighths, but how would you cut it into ten slices?

As well as the lack of decimalisation, anti-metric campaigners argued then, as now, that imperial measures are superior because they correspond to the human body. Our bodies are our first template for units of measurements – accessible, proportional, and consistent – and in nineteenth-century America and Britain, anyone who wanted to know how long an inch, foot, or span was could estimate from the width of their thumb, the length of their foot, and the spread of their arms. Such measures weren't precise, but they were always at hand. It's hard to quite imagine the utility these units provided more than a century ago, but the prospect of losing them might have felt like having a useful tool prised from your belt.

For the Institute, this spoke to the class element of their argument. They claimed to represent the common people, while those pushing metric measures were presented as faceless bureaucrats or a greedy business elite. One notable report that circulated in the US at this time came from Coleman Sellers II, an influential engineer who was president of the American Society of Mechanical Engineers. Sellers made a number of arguments against conversion, including that pre-metric France was in greater need of reform than America and that the cost of refitting American tool factories would be prohibitive. But most importantly, he says, America's measures work for those who *generate* the country's wealth, rather than those who merely calculate it. As Sellers writes: 'To the teacher, to the closet scholar, to the professional man, to those who never handled a rule or measure, but only use weights and measures in calculation, it may seem merely a matter of legal enactment; but to the worker, the dealers in the market places, to those who produce the wealth and prosperity of the land, the question is a most serious one.'[41]

Smyth himself raged on this topic with the hypnotic repetitions of a preacher. Those pushing for metric conversion, he said, were 'hasting to be rich – very rich. Rich already, as they are, they seek to add still more riches unto riches.'[42] He contrasted the idle wealth of merchants with the lives of labourers who would be most affected by the changes. 'Practical weights and measures are primarily affairs of the working classes; of the poor, and those who with their own hands do the daily work of the world,' he wrote, 'not of the rich, who lazily drink in the mere fruits and quintessence thereof without toil to themselves.' We can see how the nineteenth-century Battle of the Standards resembles present-day culture wars, pitching an 'authentic' class of workers against an out-of-touch elite. Considering the origin of the metric system came from the minds of France's *savants*, it's not a wholly unfair contrast either.

What really saved imperial measures in the nineteenth century, though, was more brutish than any argument concerning faith or tradition. It was the simple fact that Great Britain and America didn't *need* to change. The UK, for much of the century, was the world's foremost industrial nation, before being overtaken by America, and a huge quantity of both trade and industrial capital was locked into existing systems of measurement. Changing horses mid-race did not seem like the best solution to many politicians. In 1877, a resolution in the US House of Representatives was introduced, asking the executive departments of government to see 'what objections, if any' they had to making the metric system obligatory for all government transactions. All responses but one came back negative, with answers citing the familiarity of US customary measures and the cost and effort of switching to metric.[43] In the UK, meanwhile, metric measures were given equal legal status to imperial in 1896, and full adoption might have eventually followed if it wasn't for the advent of two world wars and the economic devastation that followed. Those were events even Piazzi Smyth's pyramid-inches couldn't predict.

Tilting at signposts

About a third of the way into George Orwell's *1984*, our protagonist, Winston Smith, wanders into a pub in search of evidence about life in Airstrip One before the Party took power. He corners an old man to interrogate him about this history, but finds that the prole's memories are 'nothing but a rubbish-heap of details'. To Winston's disappointment, this living link to the past is less concerned with the ascendancy of a totalitarian state than the disappearance of his familiar comforts – most notably a pint of beer, which he cannot persuade the barman to serve him.

'I arst you civil enough, didn't I?' said the old man, straightening his shoulders pugnaciously. 'You telling me you ain't got a pint mug in the 'ole bleeding boozer?'

'And what in hell's name IS a pint?' said the barman, leaning forward with the tips of his fingers on the counter.

''Ark at 'im! Calls 'isself a barman and don't know what a pint is! Why, a pint's the 'alf of a quart, and there's four quarts to the gallon. 'Ave to teach you the A, B, C next.'[44]

The missing pints signify the destruction of Britain's traditional culture and the Party's total control over everyday life. Like Newspeak, the simplified language designed to make subversive thought impossible, the imposition of metric measures is intended to effect a break with the past: destroying the old 'natural' order to fit a bureaucrat's idea of efficiency. Indeed, after the old man finally receives a metric drink, he complains that it's just not the right size for human use: "'A 'alf litre ain't enough. It don't satisfy. And a 'ole litre's too much. It starts my bladder running. Let alone the price.'"

It's this bit of Orwell that springs to mind in the pub in Thaxted, as Tony and I drain the last of our own drinks and head out to his parked car. His is a quixotic mission, a doomed last stand against the apparently unstoppable forces of modernity. But it's one that resonates with many, rooted in a love of tradition, community, and place.

As Fanny Trollope lamented of the inhabitants of the United States in the previous century: 'No village bell ever summoned them to prayer ... When they die, no spot sacred by ancient reverence will receive their bones.' For Tony, a similar loss is taking place as imperial units are removed from the country.

We've spent the morning preparing for the raid by identifying the signposts that need changing. We paced various paths around the village to check the distances in imperial (one pace is one yard, Tony reminds me) and assembled the labels that will be used to cover up the metric units. Now, we're ready, on as near as possible to a war footing as one can be on the metrological battlefield. From the boot of his car, Tony hands me my armour of officialdom: a high-vis vest with clipboard and binder. Ludicrously, I find that I'm breathing quite hard. All we're doing is putting stickers on signposts, but it still feels like – it still *is* – an illicit activity. What if someone stops us? What if the police arrive? Tony himself has been arrested for his work, even spending a night in the cells in 2001 after police officers in Kent objected to his abduction of dozens of road signs. He was found guilty of theft and criminal damage and sentenced to fifty hours of community service for each offence. A panel of judges later overturned the theft charge (though retained the one for criminal damage), ruling that there was no evidence that Tony planned to actually destroy the signs.[45]

'Do you ever feel nervous doing this sort of thing?' I ask, as he shoulders a stepladder and grabs a tube of industrial glue, checking the nozzle is clear with the air of a soldier inspecting his gun. 'Oh yes,' he says. 'Every time. But that's part of it.'

Mostly, he says, people don't bat an eyelid when ARM members are working. The high-vis vests see to that. Earlier on, in a different pub, he'd shown me a video demonstrating his point. It's called 'Uniform Obedience' and features an actor wearing a nondescript but official-looking outfit, standing in a city centre and asking members of the public to perform increasingly bizarre tasks. 'Would you just walk to the left of that apple please, sir?' he asks one passer-by, indicating a core on the pavement. 'That's right, to the left. Now, could you also stamp on that paving stone please? Thank you so much. Just to test the weight, you see.' Everyone in the video does what the man says (or at least, everyone the directors decided to show), and for Tony this illustrates something fundamental about human nature, about our unthinking obedience to arbitrary rules. I don't ask if he sees himself as an exception to this, but assume he does. Instead, I tell him that Banksy also dresses up as a council worker to get away with his own public amendments, and Tony is tickled by the comparison. 'Does he really now? That's fascinating, just fascinating,' he laughs.

On the odd occasions when members of the public do take notice of ARM's work, he says, they're either happy to see the signs being changed or dismiss the work as council pedantry. 'I remember once in Ely, there was a group of us out changing signs, and a couple stopped to watch us for a while.' The ARM members thought they might be in trouble, he says, but after a few minutes the strangers moved on. 'And as they started walking away, we heard them say, "Typical council job – four of them just to do that one sign!"'

Today, there are few holdouts to the metric system, with most sources citing the US, Liberia, and Myanmar as the only 'officially' non-metric nations. But these divisions are a little false. Metrication is a process rather than an event, and metric units are often adopted piecemeal across different sectors in every country. Many industrialised countries also make concessions to tradition, retaining the names or values of old units in one form or another. In many European nations, for example, a weight of 500 grams is informally referred to by the old word for 'pound' – whether that's *livre* in France, *Pfund* in Germany, or *libbra* in Italy.

The UK and US both seem stuck in their compromises. (They are, after all, 'the only nations called "brethren" upon the face of the Earth', as the pyramidologist Totten put it.[46]) In the UK, there's widespread support for the retention of imperial measures, at least in certain areas of life. Polls show no one wants to give up pints in the pub or miles on the road signs, though, as you might expect, that support is waning: younger generations are increasingly familiar with metric and open to its adoption in imperial fiefdoms. Brexit gave hope to some that imperial could make a comeback: a poll of Leave voters found that half wanted to bring back pounds and ounces in stores (the same proportion was in favour of the return of the death penalty[47]). But this would be horrifically impractical, overturning decades of careful work for the sake of appeasing a disproportionately vocal section of the public. Even for a movement like Brexit, happy to inflict economic self-harm for the sake of perceived sovereignty, it seems like a nuisance too far.

In the US, the last serious push for metrication came in the 1970s but met the same resistance it had in the nineteenth century, with opponents emphasising the threat of foreign ideas, the harm to the common worker, and the superiority of America's 'natural' units. Work on metrication began in 1975, when President Gerald Ford signed the Metric Conversion Act, establishing a 'national policy of coordinating and planning for the increased use of the metric system.'[48] A huge propaganda campaign followed, with Saturday-morning cartoons, infomercials, and posters proclaiming that the metric system was really happening. Certain industries, like auto-making, took the plunge and switched systems, but a lack of mandatory enforcement meant the campaign fizzled out.

That was when the culture warriors moved in. 'Metric is

definitely communist,' said Dean Krakel, director of the National Cowboy Hall of Fame in Oklahoma. 'One monetary system, one language, one weight and measurement system, one world – all communist.' Syndicated *Chicago Tribune* columnist Bob Greene dedicated several columns to his metric grievances, decrying its imposition on the US as a waste of government money and the fault of some Arabs, 'with some Frenchies and Limeys thrown in.'[49] Stewart Brand, publisher of the *Whole Earth Catalog*, a counter-culture magazine that influenced the likes of Apple founder Steve Jobs, wrote in *New Scientist* about the virtues of measures based on the human body. 'The genius of customary measure is its highly evolved sophistication in terms of use by hand and eye,' wrote Brand. 'Metric works fine on paper (and in school) where it is basically counting, but when you try to cook, carpenter, or shop with it, metric fights your hand.' He said the only reason there wasn't more of an outcry about metric was that 'the whole scheme was never taken seriously.'[50]

The zenith of America's anti-metric campaign came in 1981 with the 'Foot Ball', a metrological gala for the New York elite. As *The New York Times* reported, some eight hundred guests attended the event, where 'East Side fashion and Greenwich Village punk' mixed on the dance floor, attendees wore badges reading 'Don't give an inch', and the evening's entertainment included a Most Beautiful Foot contest. The author and inveterate contrarian Tom Wolfe attended in his trademark white suit, providing some intellectual elan alongside the necessary newspaper-ready quotes. 'I hear that the meter is based on a rod somewhere outside of Paris,' Wolfe told the *Times*. 'To use that as a basis for measurement is completely arbitrary and intellectual. I should say I have tremendous admiration for the French, but a matter of this importance should not be left to them.'[51] The following year, opinion polls showed that the majority of Americans were opposed to the adoption of the metric units, and the project was happily scrapped by President Ronald Reagan, another scalp in his administration's programme of budget cuts.

Despite this official rebuff, the US is definitely much more metric than it first appears. After all, the federal government has relied on metric units to define feet, pounds, and ounces since 1893,[52] judging metric standards to be the product of an unimpeachably rigorous scientific process. Many commercial products in the US list measurements in both metric and customary units, the better to appeal to international markets; numerous industries are metric, like automaking and pharmaceuticals; and the US military is mostly metric, to better work alongside international forces. Nevertheless, since the 1970s any legislative attempts to go metric have failed conclusively, and it's clear that cultural objections are as potent as ever. In a segment on Fox News in 2019, the notoriously xenophobic right-wing host Tucker Carlson and his guest, *New Criterion* editor James Panero, romped through some of the greatest hits of the anti-metric brigade. Carlson derided metric as a 'weird, utopian, inelegant, creepy system that we alone have resisted', while Panero praised customary measures for their derivation from 'ancient knowledge, ancient wisdom'. A scrolling ticker at the bottom of the screen posed the sort of purposefully thoughtless query typical of contemporary political trolling. 'Is the metric system completely made up?' it asked.[53] Well, yes, is the only answer, but what isn't?

———

Back in Thaxted, these arguments seem immaterial. What matters is the rickety, paint-spattered stepladder that Tony is placing in front of a green and gold signpost. It wobbles perilously on the paving stones, but Tony is confident. In just a few seconds, he's laid out the new imperial signs, squirted glue on to their backs, and ascended to press them firmly on to the signpost's fingers. I snap a few pictures, then check the imaginary paperwork on my clipboard, nervously glancing over my shoulder to spot some of the earlier locals now staring at us out of the pub window. High-vis vests are decent disguises, but they don't make you unrecognisable. What did they think we were doing?

As I worry, Tony works, swiftly gluing and placing more imperial units on to the signpost. In a trice he's done one, two, eight, and ten. The practice shows. After the last sticker is up, he swiftly kicks his stepladder into order and grabs his bag. 'Come on,' he tells me, 'there's another down the road we can do.' As he marches away, I look back at his handiwork: 'Windmill, 240 yards', 'Almshouses, 540 yards'. It's tidy, as far as unauthorised amendment to street signage goes, but also very obviously a bit of plastic stuck on a signpost. The next cluster of signs are dispatched with similar speed, and I continue in my now established role of being worried and useless. Five minutes later, the raid is over. We trot back to the car, shove our gear into the boot and whip off our high-vis vests like bank robbers pulling off balaclavas. Tony is glowing and triumphant. I'm elated, too, despite myself. ARM has struck again.

We manoeuvre out of the car park, and I have one last go at trying to crack the puzzle of Tony's anti-metric crusade. Why is he doing this, *really*? Is it love of tradition and England? Is it a religiously inspired crusade? Or is he just bored? He declines to answer the question directly, but admits that changing signposts in quaint rural towns is not really his priority in life any more. 'As my Christian faith has grown, I've become more interested in living that,' he says. 'So this is a residual activity for me. We've got a few people who are bold enough to do it themselves, but not many.' Really, it seems like the fight for traditional weights and measures is over, at least for Tony and ARM. I've found myself beguiled by the arguments of these traditionalists; by the satisfying historical and cultural density of older measures, and the admirable desire to retain their legacy in an increasingly abstracted world. But although these units once embodied important realities of every-day life, these aspects of their use are increasingly irrelevant. For example, although it's true that base-12 and base-16 divisions of imperial units make dividing goods by halves, thirds, and quar-ters easier, of what relevance is that in a world of pre-packaged groceries? And while we praise older units for being built on a more 'human' scale, is there anything more human than reach-ing beyond our grasp? To do so is a defining characteristic of the modern world, which encompasses spans beyond the individual's comprehension. As anti-metric advocates love to point out, what ultimately determines the 'right' measurement is familiarity and tradition. But tradition is not immune to change, and if imperial measures are abandoned because they're no longer useful, then that is natural too.

As we wind our way through the country back roads to the near-est station, chatting about what have become familiar topics – past ARM exploits, the necessity of protest, and the roots of English culture – Tony notes that Thaxted was once the home of English composer Gustav Holst, who worked on his famous *Planets* suite here. Holst would later adapt the main theme from the suite's 'Jupiter' movement as the hymn tune 'Thaxted', fitting it to the patriotic poem 'I Vow to Thee, My Country', which has become a staple of the UK's most prominent nationalistic events: the funer-als of prime ministers, Church of England services, and the BBC Proms. 'Ideally,' says Tony, 'I would have changed the sign with that

9

FOR ALL TIMES, FOR ALL PEOPLE

———————

*How metric units transcended physical reality
and conquered the world*

The laws of nature will defend themselves – but error – (he would add, looking earnestly at my mother) – error, Sir, creeps in thro' the minute holes and small crevices which human nature leaves unguarded.

—LAURENCE STERNE, *TRISTRAM SHANDY*[1]

The march to the heavens

On a damp but cheerful Friday in November 2018, I travelled to the outskirts of Paris to witness the overthrow of a king. I was attending events as a journalist, and spent the morning eavesdropping on scientists and diplomats from dozens of nations as they sipped coffee and gossiped about the 'biggest revolution in measurement since the French one'. The meeting was taking place in a theatre not far from the Palace of Versailles, the château where Louis XVI had reigned as the embodiment of the state until the *sans-culottes* came for his head. But today matters would be settled by a vote.

The goal of the attendees was to dethrone the last remaining physical standard of the metric system: the kilogram. Since 1799, this unit of weight has been defined using a lump of metal stored in Paris, with the current office holder – the International Prototype Kilogram, or IPK – installed in 1889. As we mingle in the lobby, the IPK, affectionately known as *Le Grand K*, is locked up just a few miles away, sealed in a triple-layer bell jar in an underground vault. It is *the* kilogram: what it weighs, every kilogram in the world weighs, no more and no less. Chip a piece off its sides and every set of scales in the world needs recalibrating. Which is why it has to go.

If you were to break into the IPK's vault, you would find underneath the bell jars a small, bevelled cylinder with an unmarked exterior polished to a mirror shine. Its design echoes the Jacobin plainness of the original metre, and it is similarly cast from the most enduring material of its time: a platinum–iridium alloy specially selected for its metallurgical stability. A side effect of its composition is that the kilogram is extremely dense and therefore menacingly compact. If you picked it up – which you're not allowed to – it would nestle in the palm of your hand no bigger than a hen's egg, exerting a downward force disproportionate to its size. Most of us are simply not used to handling objects of such density and, as a result, the IPK seems to tug at your hand, as if Aristotle were right all along about the soul-like purpose of matter, and it was straining to return to the earth. As one former keeper of the kilogram told me, such weightiness can catch one off guard: 'The first time I picked it up with tongs I nearly dropped it,' he confided.

Physical standards like this kilogram weight owned by NIST are still used in metrology to promulgate exact values of units

From its underground vault in Paris, the kilogram issues metrological orders around the world. Copies are made by the International Bureau of Weights and Measures, or BIPM. This body is in charge of the metric system, formally known as the SI or International System of Units, and sends copies of the kilogram to labs around the world, who use them to validate commercial weights and calibrate scales for their clients. These, in turn, are then used to weigh everything from iron ore to apricots, the validity of each measure traceable back through a labyrinth of commerce and industry to the vault where the kilogram resides. Like the Holy Ghost, the IPK is singular but its presence is multiple, and whether you are measuring flour for bread, lifting weights in the gym, or buying timber by the tonne, it is *Le Grand K* that sits, invisible, on the other side of the scales. Even nominally non-metric countries like the US define their units using metric standards, and in the privacy of laboratories and factories they too bow to the all-conquering power of the metric system. It's the sort of absolute authority that would have made Louis XVI blush.

On this damp Friday in Paris, though, the IPK is not long for this world. As guests and media mingle in a foyer adorned with metrological artefacts and metric-themed cakes, I spot a familiar and welcome sight: the irrepressible grin and round, thick-rimmed glasses of Stephan Schlamminger, a physicist at America's National Institute of Standards and Technology (NIST), the country's home of measurement.

Schlamminger is something of a *genius loci* of metrology: an animating spirit full of cheer and knowledge, as comfortable in the world of weights and measures as a fire in a hearth. He is also a key player in the American team that helped create the kilogram's new definition. I'd spoken to him before, and always delighted in his enthusiasm and generosity. 'James, James, James,' he says in a rapid-fire German accent as he beckoned me to join his group. 'Welcome to the party.'

As Schlamminger tells it, the redefinition of the kilogram is the fulfilment of the historical arc that began with the French Revolution. The goal of the modern metrologist, he says, is the same as that of the eighteenth-century *savant*: to create measurements that are *à tous les temps, à tous les peuples* – for all times and for all people. To emphasise the point, he rolls up his shirt sleeve to reveal these words tattooed on his forearm, a tribute shared with the other members of his team. 'We said if we pulled this off, we'd all get tattoos,' he tells me. 'I finally got mine two weeks ago.' The motto reflects the aspirations of metrology – of durability, precision, and replication – but also hints at its underlying challenge: that science never stands still. The devil is in the decimals, and as scientific instrumentation has become more precise through the centuries, the demands of standards increase in tow. The problem of the eighteenth-century astronomers, whose improved telescopes discovered more error, never really goes away.

The first metric units met these demands by taking their values 'from nature',[2] with the metre defined as a fraction of the Earth's meridian and the kilogram as the weight of a cubic decimetre of water. But once these values were established, the units themselves needed to be made accessible, and for that you need physical standards, with all their flaws. At more than a century old, the current International Prototype Kilogram has lasted well, but problems have emerged all the same. The most important: it has lost weight. This discrepancy was discovered during one of its semi-regular weigh-ins – an event that takes place every forty years or so, where national standards from around the world are flown into Paris to be compared with *Le Grand K* and its honour guard: a set of six *témoin*, or 'witness', kilograms that were cast at the same time as the IPK and are stored in the vault alongside it.

These weigh-ins resemble the treatment of grain measures in medieval Europe, with every movement scrutinised and every variable controlled. The end effect turns protocol into ritual. The sacral objects are the standards themselves, which have to be scrupulously cleaned before being weighed. Each one is rubbed down by hand with a chamois leather soaked in a mixture of ether and ethanol and steam washed with twice-distilled water. Given the high stakes of the measurement, absolutely nothing is left to chance, with the BIPM's official cleaning manual describing every step in meticulous detail, from the amount of pressure to be applied with the chamois (around 10 kilopascals) to the distance between kilogram and steam-cleaner (5 millimetres). Even the method of removing excess water using filter paper is carefully described: 'For this operation, an edge of the paper is put in contact with each drop and the water allowed to flow into the paper by capillary action.'[3] It is a secular sacrament, designed to appease the gods of metrology and maintain the reputation of international measurement, a system that supports so much in the modern world.

Despite these ministrations – or rather, because of them – it was discovered in 1988 that the mass of the IPK was diverging from that of the *témoins* and national standards. They were getting heavier and it was getting lighter. Not by much, certainly: the difference between the kilogram standards averaged out to just 50 micrograms, or the mass of a fingerprint.[4] It would be unnoticeable in any other circumstance, but was grave cause for concern here. If the weights were diverging, which was the correct value? How do you verify the mass of an object that itself defines mass? By global agreement, whatever the IPK weighs is a kilogram. That means it can't lose weight – everything else in the universe just gets a tiny bit heavier instead.

Luckily, there is a solution. The kilogram isn't the first standard to reveal such variation under scrutiny, and over the last century the

BIPM has in fact rid itself of every other physical artefact that supports the metric system. One by one, these pillars of measurement have been replaced by definitions not beholden to the material world at all. Instead of tying units to physical standards, the metric system is now defined using fundamental constants of nature, properties we believe to be inherent to reality. The metre, for example, is no longer equal to the length of a metal bar but to the distance travelled by light in 1/299,792,458th of a second. The second, in turn, is no longer 1/86,400th of a solar day but the duration of 9,192,631,770 radioactive cycles of an atom of caesium-133.[5] Physical standards are still used to share these measures around the world, but, in theory, any laboratory in the world can define the units of the metric system from scratch.

In Versailles, the prospect of ridding the world of its last physical standard makes Schlamminger's eyes gleam in excitement. 'As long as they define the kilogram as an artefact,' he tells me, 'we cannot say this is for all people, for all times. It is not "for all people" because people cannot remake the IPK, and it is not "for all times" because it is an object and every object changes. None are immutable.' That's why everyone has gathered in Paris: to root metrology in something deeper and more permanent. Schlamminger is practically singing as he contemplates the coming change. 'I really can't believe it, that today is the day! It's unbelievable,' he says. 'We will transcend this messiness. We will be basing units on the fabric of the universe: on the heavens, so to speak!'[6]

Charles Sanders Peirce: 'manic grandiosity' and an 'unbearable sense of futility'

The march to the heavens is long, though, and the quest to define units of measurement using constants of nature began more than 150 years ago, with a man named Charles Sanders Peirce. Born in

Cambridge, Massachusetts, in 1839, Peirce (pronounced 'purse') is not a well-known figure in intellectual history, but for those familiar with his work no praise is high enough. To admirers he is 'the American Aristotle', the continent's 'most original and most versatile intellect', and creator of a body of work 'without precedent in the history of the Earth'.[7] He was best known as a philosopher and logician, but produced work as a chemist, engineer, psychologist, astronomer, mathematician, and, of course, a metrologist. His intellectual range was matched by a personal inconstancy, which hampered his professional ambitions. He was an insomniac given to bouts of 'manic grandiosity and visionary expansiveness'; a depressive periodically laid low by an 'unbearable sense of futility';[8] and a philanderer whose abrasive character would alienate loved ones and galvanise his enemies. He was, to lean on the powers of understatement for a moment, a complicated man.

In contrast to his personal tumult, the questions that motivated Peirce were simplicity itself. 'What do we know and how do we know that we know it?' were his main concerns, which he answered, in part, by creating the first philosophical tradition the United States can truly call its own: pragmatism. Like any school of philosophy, pragmatism has its shades and subtleties, but is best encapsulated by the idea that the meaning of things can be discovered from their effect on the world. As Peirce's friend and fellow pragmatist, psychologist William James, put it: 'The ultimate test for us of what a truth means is indeed the conduct it dictates or inspires.'[9]

When applied to Peirce's own life, this thesis has mixed results. His thinking had a great influence on society, but mainly when modulated by the work of others. He never found much academic acclaim and would die, isolated and penniless, convinced that he would yet produce the work that would shake the world. As he reflected on his life as an old man, he decided that his fatal flaw had been a lack of

A mathematician, philosopher, and metrologist, Charles Sanders Peirce
was the first to experimentally tie a unit of length to a constant of nature

self-control: 'For long years I suffered unspeakably, being an excessively emotional fellow, from ignorance of how to go to work to
acquire a sovereignty over myself.'[10]

It's a dictate that can be traced back to Peirce's father Benjamin,
a legendary figure in his own right who helped found the National
Academy of Sciences under President Lincoln. The elder Peirce recognised his son's potential from a young age and burdened him with
the label of 'genius' before he reached adolescence. He schooled
Peirce in maths, logic, and philosophy, but above all sought to teach
him the power of attention, training him with marathon sessions of
card problems that could last from the evening until sunrise, accompanied by sharp criticism if he made any mistakes.[11] Peirce idolised
his father and internalised these lessons, writing that if he ever had a
son (he never did), he would teach him 'there is but one thing that
raises one individual animal above another, – Self-Mastery'.[12]

284

This quality would be particularly difficult for Peirce to attain, as he developed a neurological condition as a young man that would plague him all his life: trigeminal neuralgia, which assails the victim with random and excruciating bursts of pain in the face and jaw. Peirce treated the condition with a cocktail of opium, morphine, ether, and cocaine – all legal prescriptions at the time, though he took them to excess.

The unpredictable nature of his condition seems to have been like a shock collar fitted to an uncomprehending animal. Biographers have speculated that it provoked the personality swings that alienated would-be admirers and instilled in Peirce a relentless work ethic. Every pain-free moment was a gift that had to be used to write and think. As a consequence, Peirce left behind more than 100,000 manuscript pages, and though he published only a single book in his lifetime (on the measurement of light from stars), he was convinced that he would create a philosophical system 'so comprehensive that, for a long time to come, the entire work of human reason ... shall appear as the filling up of its details'.[13]

The romance of the next decimal place

Peirce's introduction to metrology came with his first job: at the US Coast Survey. This was a scientific agency established by Thomas Jefferson in 1807 to define the physical boundaries of America and was, consequently, enmeshed in the world of measurement. Peirce would work at the Survey for more than thirty years, moving through different roles, including becoming head of the agency's Office of Weights and Measures. And although he considered this work a distraction from logic and semiotics, it played a crucial role in developing his philosophy of science.

It was a good time to be working in measurement, with precision a byword of both science and industry. The term 'metrology' first appears in the English language in 1816,[14] and the field underpinned practical inventions like the railway and telegraph, as well as theoretical breakthroughs like thermodynamics and electromagnetism. Looking back on the century in 1931, the physicist Floyd K. Richtmyer noted that what the printing press had done for the medieval mind, measurement did for nineteenth-century science. '[The] outstanding characteristic of nineteenth century physics ... is the extent to which the making of precise measurements, merely for the sake of securing data of greater accuracy, became a recognised part of research in physical laboratories,' he wrote in *Science* magazine, in a piece titled 'The Romance of the Next Decimal Place'. 'From each extension of the precision of measurement there results either a significant modification of theory, or not infrequently a new discovery ... So frequently has this happened ... I am disposed to conclude by paraphrasing a famous saying: "Look after the next decimal place and physical theories will take care of themselves."'[15]

The measurement of electricity offers a clear example of this, as work that not only involved theoretical breakthroughs and fascinated some of the greatest minds of the era (including James Clerk Maxwell, Michael Faraday, and Lord Kelvin), but was also bound up in commercial and political demands. For electricity to be bought and sold it had to be measured accurately, while standardising units of electrical resistance was crucial to the creation of telegraph lines. It was particularly needed for maintenance, as a technician who knew the length of a given cable and was able to measure its electrical resistance accurately could quickly locate a fault with a bit of basic maths. Without these calculations, technicians had to dig or take down hundreds of metres of cable. Keeping these lines crackling with messages was essential not just for businesses but for political control too. In the British Raj, for example, the power of the telegraph allowed a relatively small number of troops (roughly 66,000 British soldiers and 130,000 Indians) to exert control over a population of more than 250 million.[16] After the Indian Rebellion of 1857 failed, British newspapers credited victory to British technology. 'Never since its discovery has the electric telegraph played so important and daring a role as it does now in India; without it, the Commander-in-Chief would lose the effect of half his force,' wrote a correspondent for *The Times*, while one British official later declared: 'The Electric Telegraph has saved India.'[17]

Not coincidentally, the first half of the nineteenth century saw the consolidation of the world's two great rival systems of measurement. The metric system had failed on first contact in France, but the principles of its creation – scientific neutrality and equality – aided its spread across Europe. Partly in response to the rise of metric measures, the UK overhauled its own measurement system, which dated back to the Romans and Anglo-Saxons. In 1824, it created the British Imperial System, which had its authority guaranteed not by its derivation from the Earth, but by the military and economic might of the largest empire in history; though, learning from the French example, the British government went to great efforts to create accurate physical standards that could be shared around the world. Just because the value of yards, feet, and inches had their origin in history rather than science, that didn't mean they could be inconsistent.

All of this attention and energy lofted metrology in the nineteenth century to 'a plateau unheard of in previous eras', reaching new standards of 'reliability, accuracy, precision, and durability'.[18] But while these systems were certainly good enough for trade and industry, their shortcomings were becoming more noticeable in the lab. This was because measurement remained a primitive field in one crucial regard. Just like Egyptian pharaohs and Mesopotamian kings, the chemists, engineers, and cartographers of the nineteenth century relied on physical standards in order to weigh and measure the world. No matter how much intellect went into the creation of the metre or yard, the value of these units was still defined by lumps of metal, as susceptible to decay as any matter on Earth.

The danger of this situation was demonstrated with dramatic emphasis in 1834, when the UK's seat of parliament, the Old Palace of Westminster, burned down and took with it the country's standard yard and pound. Ironically, the fire itself was caused by the disposal of another ancient tool of reckoning: tallies. These are short lengths of wood carved with notches to represent money owed. These staves are then split down their length into two pieces, foil and stock, which are given to the debtor and creditor respectively. The unique shape of the wood's split ensures this record cannot be forged (and is also where the term 'stockholder' originates). The British government had been using tally sticks in its accounting since the medieval era, but had finally decided to get rid of these old records. Two cartloads of tallies were burned in furnaces in the palace's basement, but the fire spread and engulfed the building, taking with it the country's standards of measurement.[19]

In 1870, James Clerk Maxwell outlined the problem clearly. The century before, the creators of the metric system had based their units on the most stable foundation they knew: the planet under their feet. But this was short-sighted, he said. The Earth might 'contract by cooling, or it might be enlarged by a layer of meteorites falling on it, or its rate of revolution might slowly slacken'. Instead, he said, scientists should look to the century's newly discovered territories: 'If we wish to obtain standards of length, time and mass which shall be absolutely permanent, we must seek them not in the dimensions or the motion, or the mass of our planet, but in the wavelength, the period of vibration, and the absolute mass of these imperishable and unalterable and perfectly similar molecules.'[20]

Maxwell's suggestion was sound, but he did not offer a practical way to define the metre in this fashion. Indeed, many scientists in this period were sceptical of such a basis for measurement. Patrick Kelly, the mathematician who coined the word 'metrology', worried that 'nature seems to refuse invariable standards: for, as science advances, difficulties are found to multiply, or at least, they become more perceptible, and some appear insuperable'.[21] And when the UK replaced its measurement standards in 1834, the astronomer John Herschel, who was charged with creating the standards, made a point of emphasising the arbitrary nature of the standard and the painstaking work of its creation. 'Our yard', he wrote, 'is a purely individual material object, multiplied and perpetuated by careful copying; and from which all reference to a natural origin is studiously excluded, as much as if it had dropped from the clouds.'[22] Instead of turning to scientific abstracts for authority, the British preferred to rely on 'law and labor'.[23]

For Peirce, though, the challenge of defining units of measurement using constants of nature was energising. He was already aware of the fallibility of physical standards after taking charge of the Coast Survey's department of Pendulum Experiments, which had led him to discover that the physical standards enshrined by the US government varied worryingly, both from one another and their European counterparts.[24] As he later wrote: 'The non-scientific mind has the most ridiculous ideas of the precision of laboratory work, and would be much surprised to learn that, excepting electrical measurements, the bulk of it does not exceed the precision of an upholsterer who comes to measure a window for a pair of curtains.'[25]

To aid the general development of US metrology, Peirce began travelling to France, Germany, and Britain to learn from the experts there. Between 1870 and 1883, he made a total of five trips, shuttling standards back and forth and meeting with some of the foremost scientists of the day, including Maxwell. The writer Henry James recalls meeting Peirce for dinner in Paris and gives a somewhat typical account of him in letters to James's brother William. Peirce, he says, lived 'a life of insupportable loneliness and sterility but of much material luxury. He sees, literally, not a soul but myself and his secretary. [He's] a very good fellow when he is not in an ill humor; then he is intolerable.'[26]

Finding new foundations

Imagine, then, Peirce sitting in the lounge of a transatlantic ocean liner, a bundle of papers in front of him covering the latest advances in geodesy and a leather satchel at his side containing a metal length standard: all the tools one needed to measure the world. Even the ships he travelled in were proof of measurement's ability to smooth the irregularities of life. A journey from America to Europe the previous century took months, with sailing ships buffeted off course by currents and wind. But the iron-hulled ocean liners of the nineteenth century offered not only first-class accommodation and fine dining, but the greatest amenity of all: a reliable timetable for departures and arrivals. You could book a ticket in advance and be sure of the hour you would step on to land, thanks to the ships' precision-engineered steam engines.

It was during this decade of travel that Peirce began formulating his idea to derive the length of the metre from the source of universal constancy suggested by Maxwell: the frequency of light itself. His method relied on a tool known as a diffraction grating: a small piece of glass or metal inscribed with thousands of densely packed lines that scatter light into its constituent wavelengths. If you've ever looked at the back of a CD or DVD and noticed a rainbow sheen playing over the surface, you've seen a diffraction grating in action.

The tiny laser-etched ridges and valleys that encode data on the back of a CD are functionally identical to the slits or grooves used in scientists' tools, though the latter are more precisely arranged. The gratings used by Peirce were cutting-edge technology: drawn by a diamond stylus driven by an intricately geared, water-powered device – hand-crafted but capable of machine regularity, and able to rule 6,808 equally spaced and parallel lines into the width of a single centimetre.[27]

If you shine ordinary light at a grating, as with a CD held under a lamp, you'll see it splits into different colours, each corresponding to a specific wavelength in the electromagnetic spectrum. But shine a pure light source at it and you'll see only a pattern of dots corresponding to its wavelength. This pattern, constant so long as the light source is the same, is what Peirce wanted to use to set his length standard. He would record the angles of light split through his diffraction gratings to measure its wavelength and multiply that to recreate the metre.

It's a disarmingly simple method, and Peirce thought he only needed to work on the accuracy, reducing impurity in his light source and inconsistencies in the diffraction gratings. 'Metallic bars, hitherto our ultimate standards of length, probably change in length in the course of years,' he wrote in a report to the US Coast Survey in 1881. 'The confusion into which such spontaneous changes in standards of length may throw all precise measurements referred to them, is too obvious to be insisted upon.' With his new definition, though, the metre would be defined by light itself, 'and this we have reason to believe the most unalterable thing in nature'.[28]

———

Peirce's work is interesting not just because of his practical endeavours, but because of the philosophy of truth and science that he developed at the same time. It is fair to say that no other dedicated metrologist has had such an impact on philosophy as Peirce, and his thinking illuminates the drive and demands of the field perfectly. It was during his Atlantic crossing that Peirce laid the foundation stones for the philosophy of pragmatism, translating an essay originally written in French, 'The Fixation of Belief', into English and penning its follow-up, 'How to Make Our Ideas Clear'. He would later follow these with four more essays, which would collectively be titled 'Illustrations of the Logic of Science' and published in *Popular Science Monthly*.

In them, Peirce argues that the 'method of science' is the only secure path to knowledge in the world. This may seem like arrogance, but he qualifies it in a number of ways. First, he says, the scientific method is hardly the purview of scientists alone. Instead, it's practised by all enquiring minds, whether they are riddling out facts in a workshop, a crime scene, or an archaeological dig. Second, he says, this doesn't make scientists themselves infallible. They are as likely to fall prey to fallacy as any other group of humans, and he outlines a number of failings, like ignoring evidence that doesn't fit a theory or simply believing what we're told because it comes from an important authority. More often than not, Peirce thought, we fool ourselves with spurious logic and reasoning rather than relying on what can be learned from the world. Just look at all the scientific 'knowledge' of the past, from heliocentrism to caloric theory, that seemed reasonable to the mind but was disproved with facts and figures, he says. Such thinking turns enquiry into 'something similar to the development of taste',[29] Peirce argued, and all tastes go in and out of fashion.

The only thing that produces reliable data, he thought, was experiment and observation. For although Peirce was at home in the abstractions of logic and mathematics, he was adamant that truth was something produced with the hands and the eyes. He praised scientists like Lavoisier, who, in disproving the existence of phlogiston through careful measurement, had turned scales and alembics into 'instruments of thought'. In doing so, Lavoisier created a 'new conception of reasoning as something which was to be done with one's eyes open, by manipulating real things instead of words and fancies', said Peirce.[30] Notably, he also stressed that such work only succeeds if verified by others. '[O]ne man's experience is nothing if it stands alone,' he wrote. 'If he sees what others cannot see, we call it a hallucination.' It was only by working as part of a group, by collating and comparing results, that one is able to 'grind off the arbitrary and the individualistic character of thought'.[31]

In a lecture series given in 1903, Peirce illustrates his desire for description and accuracy with exhaustive flair:

If you look into a textbook of chemistry for a definition of lithium, you may be told that it is that element whose atomic weight is 7 very nearly. But if the author has a more logical mind he will tell you that if you search among minerals that are vitreous, translucent, grey or white, very hard, brittle, and insoluble, for one which imparts a crimson tinge to an unluminous flame, this mineral being triturated with lime or witherite rats-bane, and then fused, can be partly dissolved in muriatic acid; and if this solution be evaporated, and the residue be extracted with sulphuric acid, and duly purified, it can be converted by ordinary methods into a chloride, which being obtained in the solid state, fused, and electrolyzed with half a dozen powerful cells, will yield a globule of a pinkish silvery metal that will float on gasolene; and the material of that is a specimen of lithium.[32]

One last strand, though, needs to be added to Peirce's philosophy to convey the man's thought and character. And this is perhaps his most important belief of all: that knowledge is ultimately contingent – a doctrine he dubbed 'fallibilism'.

To Peirce, fallibilism meant that there are no facts in life that are beyond doubt. Everything we believe exists with the possibility that it will be proven wrong, from the base evidence of our senses to the most elaborate, rigorously tested, and apparently flawless scientific theories. The doctrine of fallibilism is distinguished from the approach of sceptics, who claim we can never know anything for sure, with the addendum that it is quite all right to believe that we know things (indeed, it is essential to living), but we must, at the same time, leave open the possibility that we are completely, spectacularly wrong.

Peirce didn't think that this doctrine of fallibilism should be difficult or even controversial to accept. 'That we can be sure of nothing in science is an ancient truth,' he notes. But its centrality to his beliefs and character is still perhaps surprising. Like the acceptance of worldly impermanence is to a Buddhist, so fallibilism was to Peirce: a precondition for learning the truth about the world, and the grit that stimulated his soul. 'Indeed, out of contrite fallibilism, combined with a high faith in the reality of knowledge, and an intense desire to find things out, all my philosophy has always seemed to me to grow,' he wrote.[33] For a boy who was taught by his father the benefits of constant, unwavering attention, perhaps this was the truth that sustained his vigilance: the constant threat of error.

Peirce's experiences with imprecision in measurement only consolidated this view: that the world was full of inaccuracies, and that even our best attempts to establish a solid foundation for truth were fragile and liable to fail. 'We never can be absolutely sure of anything, nor can we with any probability ascertain the exact value of any measure,' he laments in one unpublished manuscript.[34] This attitude – combining an urgent desire for facts with an acceptance of their fallibility – makes Peirce something close to a patron saint of metrology; someone whose character and work captures the deep antagonism at the heart of the subject. The fact is that the more we measure, the more mistakes we find; the harder we try to bear down on the truth, the more we reveal the inadequacies of our assumptions. We push and push on the knowledge and equipment we have until finally – snap! – the pencil breaks, the glass cracks, the line tears, and we are sent spinning into the void again, hoping to rebuild what we know from scratch with our own hands and eyes.

Michelson and Morley's metre

The hard truth of Peirce's struggle to define the metre using light is that he failed. Just as he might have predicted, fallibility found him out. He shared notes and summaries of his work but never produced a definition rigorous enough for either his own satisfaction or publication. But true to his philosophy of science, his method would be picked up and improved upon by others, leading not only to a redefinition of the metre, but to a revolution in physics.

By the end of the nineteenth century, many scientists felt satisfied that they had more or less determined the basic gears and levers of the physical world.[35] Maxwell's electromagnetism had unified electricity, magnetism, and light into a single coherent system, while Newton's mechanics still held sway over the world of motion, inertia, and momentum. There were some loose ends to be tied up, sure, but nothing too arduous. As the American physicist Albert A. Michelson noted in 1894: 'It seems probable that most of the grand underlying principles have been firmly established [...] The future truths of physical science are to be looked for in the sixth place of decimals.'[36]

Heeding his own advice, Michelson set out to prove one of the era's commonly accepted truths: the existence of the 'luminiferous ether', a theoretical medium through which light was supposed to travel, just as waves of water travel through the ocean. ('Luminiferous' means 'light-bearing', while 'ether' can be traced back to ancient Greece as the rarefied medium, neither liquid nor air, that occupied the heavens.) Physicists at this time were fairly certain that light was a wave of some sort, rather than a particle, and so reasoned that it needed a medium – the ether – through which to proliferate, just as sound waves travel through air. It's exactly the sort of 'common sense' reasoning that Peirce was so wary of.

To detect the presence of the ether, Michelson and his colleague Edward Morley set out to measure its effect on the speed of light. In 1887, using a device of Michelson's invention called an interferometer, they split a beam of light and bounced it off a pair of mirrors set at right angles to one another, before recombining the divided light into a single beam. If the two beams had travelled at the same speed while divided, then their wavelengths would line up perfectly with one another, creating a pattern of light and dark 'fringes' like a series of concentric circles. But if the ether existed, then it would create a crosswise drag as the Earth orbited the sun, the so-called ether wind, which would slow down one of the two beams and disrupt the pattern. Michelson later explained the experiment to his daughter by comparing the beams of light to a pair of swimmers in a race, 'one struggling upstream and back, while the other, covering the same distance, just crosses and returns. The second swimmer will always win, if there is any current in the river.'[37]

But the experiment found nothing. It showed no current, no delay, and subsequently no ether, much to the dismay of Michelson and Morley. But disproving the existence of the ether was a hugely important discovery in its own right, one that would lead to even more significant breakthroughs, and as a result the pair's work is often referred to as 'the most famous failed experiment in history'.

Before those discoveries, though, and perhaps as consolation for their failure, Michelson and Morley repurposed their equipment to measure the metre itself, defining its length by counting the 'fringes' of light created by the interferometer. In the same year they conducted their experiments, they published a paper titled 'On a Method of Making the Wave-Length of Sodium Light the Actual and Practical Standard of Length'. In it, they credited Peirce for making 'the first actual attempt' at defining the metre using light, before noting that 'the degree of accuracy [he] attained is not much greater than one part in fifty or a hundred thousand'.[38] In the following decades, their method would be further refined as scientists discovered new ways of generating light beams with ever more regular patterns of fringes. These became the de facto standard for length measurement around the world, deployed for all manner of high-precision work. Then, at last, in 1960, the BIPM made this definition official, adopting the first ever unit based on 'natural and indestructible' standards and defining the metre as 'the length equal to 1650763.73 times the wavelengths in a vacuum' of the light emitted by a krypton lamp.[39]

The story of the metre does not end there. By disproving the existence of the ether, Michelson and Morley's experiments had created a crack in the foundations of physics. It would not be resolved until 1905, with the publication of four papers by Albert Einstein containing, among other things, his special theory of relativity and his famous equation $E = mc^2$. Einstein's theory of special relativity had two key postulates: first, that the laws of physics are unchanging, so long as observers are moving at a constant speed; and second, that the speed of light is constant, no matter how slow or fast observers are moving. Einstein's theories entangled space and time in all sorts of wild and fascinating ways, leading to a complete rethinking of those certainties of the nineteenth century. It also resolved the problematic findings of the Michelson–Morley experiment. Einstein commented later that it

was this result and its subsequent analysis that laid the foundations for his revolution in relativity: 'If the Michelson–Morley
experiment had not brought us into serious embarrassment, no
one would have regarded the relativity theory as a (halfway)
redemption.'[40]

The story of twentieth-century science often means that
Michelson's comments about the future of physics residing in the
sixth decimal place are seen as hubris, sometimes even mocked for
their lack of foresight. But he wasn't exactly wrong. The future of
physics *was* found in the decimals, it just wasn't the future anyone
expected, with an experiment that involved measuring the speed of
light preparing the ground for the greatest scientific re-evaluation of
our universe since Galileo.

This good fortune ultimately benefited metrology too. With the
speed of light established by Einstein as a constant throughout the
universe, it was possible to redefine the metre once again in 1983, fixing its value as 'the length of the path travelled by light in vacuum
during a time interval of $1/299,792,458$ of a second'.[41] And because
the speed of light can itself only be measured by reference to a unit of
length, this means that this definition fixes not only the length of the
metre, but also the *speed of light*. This might seem like circular reasoning, to say that we've defined the metre using the speed of light,
which is itself defined by the metre. But it shifts the epistemological
foundation of the unit from a metal bar stored in a vault to an intangible value that is constant throughout the universe. It is also quite an
astounding end to the metre's story. This is a unit that was imagined
more than two centuries ago as both symbol and support for a new
ideal of political equality, measured first by a pair of scientists who
took sight lines from church steeples and hilltops in the midst of war
and revolution. Now it has been remade from light itself, unchanging
and absolute.

The kilogram is dead; long live the kilogram

Standing on stage in a plush theatre in Versailles, the scientist Bill Phillips is singing metrology's praises. 'The definition of the metre is both brilliant and beautiful,' he says. And today, scientists are bringing 'this same beauty to the kilogram'.

Phillips is a Nobel Prize winner in physics and a metrologist at NIST. He's tall and jovial, with a trim white beard and portly build, like a Father Christmas impersonator in the off season. On stage, he fizzes with enjoyment and enthusiasm as he explains to the audience the sheer ridiculousness of the kilogram's current situation. 'Think about this: today in the twenty-first century the unit of mass is an artefact, a piece of metal that was made in the nineteenth century based on an object that was made in the eighteenth century,' he declares in tones of mock horror. 'This scandalous situation must be fixed!'

But how to fix it? The metre's length was removed from the material world by tying it to the speed of light. Therefore, says Phillips, the kilogram needs a constant of its own. And after decades of painstaking work by metrologists and physicists, one has been found. It's a constant that acts as a complement to the speed of light, denoted as c. 'It's the yin to the speed of light's yang,' as Schlamminger puts it. It is Planck's constant: h.

Think of it like this. The speed of light defines the universe at its largest spans. It is reality's speed limit: you cannot travel faster than the speed of light, and so you cannot transmit information beyond its reach. In other words: it cannot be *exceeded*. Planck's constant, on the other hand, helps define the lower boundaries of reality by describing the smallest action possible for elementary particles. It cannot be *subceeded*. If the speed of light rules supreme among galaxies, black holes, and the spaces between stars, then Planck's constant has for its domain atoms, electrons, and the pocketable abyss of the subatomic world.

The discovery of Planck's constant was fundamental to our understanding of quantum mechanics, the revolution that followed Michelson and Morley's famous failure. 'Quantum' here refers to the fact that reality as we know it appears to be quantised, or divided into parts. It's a simple fact but forms the foundation of all quantum theory. It means that if we zoom in as far as possible to examine the smallest phenomena in the universe – like the behaviour of photons, particles of light with zero mass – then what we observe occurs in discrete units. That is to say, if you measure the energy of photons, then your results fall into discrete categories, like rungs on a ladder, rather than being infinitely variable, as if they existed on a continuum. Planck's constant is what defines the distance *between* those rungs.

In a more practical sense, Planck's constant is also mind-bogglingly small. Indeed, it can be used to define what are theoretically the smallest length, mass, and temperature possible, values below which these qualities as we understand them simply disappear, they no longer make sense as a description of reality. The smallest length, for example, is Planck's length, which can be rendered numerically as 1.6×10^{-35} metres. To put it another way, Planck's length is so minuscule that if you scaled up an atom to the size of the Earth, the distance would still be smaller than an atom is now.[42] As far as we know, this is as small as the world gets, and this is why we talk about the quantum universe: there is a bottom to reality that cannot, apparently, be further divided. Thinking about all this is immensely difficult and counter to all our intuitions learned from the macro world, but really, understanding quantum mechanics and its implications is a little like contemplating a Buddhist koan: what is required is not intelligence per se but acceptance.

Rather unsettlingly, though, Planck's constant can still be measured. In fact, it can be measured through many different methods, which suggests, however improbably, that our value for it is the right one. It was one of these methods that was used to redefine the kilogram, with the help of a device known as a Kibble balance.

A Kibble balance looks like it was stolen from the set of a sci-fi B-movie. It looks like someone took an early steam engine, the sort that claimed the hands of Victorian orphans, and rebuilt it using steel and chrome. It looks, in fact, exactly as elaborate and impressive as it is, with a huge central wheel and two pan balances surrounded by a supporting cast of struts, wires, valves, and pipes. But while a normal balance simply weighs one object against another by balancing the force of gravity exerted on both, the Kibble balance weighs one object against an electromagnetic force, the magnitude of which is measured with extraordinary precision. Indeed, the precision required for the Kibble balance is so fine that it's operated in a vacuum and the technicians who operate it have to factor in changes in gravity caused by the moon's position as it orbits the Earth.

How exactly the balance measures the kilogram using Planck's constant is a fiddly process, but can be explained using two simple equations. The first is one we've seen already in this chapter, $E = mc^2$, where E is energy, m is mass and c is the speed of light. The second is less well known but equally fundamental to physics. It also happens to be the first proper equation used to describe the quantum world: $E = h\nu$, where E is energy, ν is frequency (it's not a 'v' but a lower-case Greek letter, nu), and h is our old friend Planck's constant.

We can see here how there is overlap between these two equations. $E = mc^2$ shows that mass can be measured in terms of energy as long as we know the speed of light, while $E = h\nu$ shows that energy can then be measured in terms of frequency, a feature of all electromagnetic waves, as long as we know h, Planck's constant. Juggle the sides of these equations around a bit and we get a new formula, $m = hf/c^2$, which allows us to define mass from these other values. This new equation is what the Kibble balance calculates by precisely measuring a kilogram's weight in terms of electrical forces. The end result is similar to fixing the metre to the speed of light. In both cases we know that the constant exists, but there is some degree of uncertainty in measuring it. However, by measuring the unit, be that kilogram or metre, we also fix the constant for future calculations. And, true to the yin and yang nature of c and h, instead of the huge figure we get for the speed of light, 299,792,458 m/s, Planck's constant has a correspondingly tiny value: $6.62607015 \times 10^{-34}$ J s, or 0.6 trillionths of a trillionth of a billionth of 1 joule-second.[43] A single joule-second, for context, is roughly the amount of energy required to lift a single apple a metre into the air. So imagine 0.6 trillionths of a trillionth of a billionth of *that*. The universe literally can't do less.

––––––––

As Phillips finishes his explanation of the kilogram's redefinition, he invites the scientists involved in the work to stand up and take a bow. He begins naming colleagues, demanding that they stand up to receive their adulation. 'The people who have been working on the Kibble balance, I want you to stand up! The people who have been working on electrical standards, I want you to stand up!' he says, as individuals and groups from labs around the world cautiously rise to their feet. 'On quantum standards: Klaus, stand up! Ulrich, stand up! All the people, all the people!' shouts Phillips. More and more metrologists are located, until a few dozen men and women are standing around the room. The applause of the audience gathers strength before rolling around the theatre like a great storm.

For outsiders like myself, it's not clear exactly what we're applauding. The challenges of modern metrology are undeniably arcane and involve the latest research in numerous fields. And as many people tell me in Versailles, if the redefinition of the kilogram goes to plan, then no one will even know it happened. Despite this, the sense of overwhelming *pride* in Versailles is irrepressible, dissolving all barriers of ignorance. As Peirce made clear in his philosophy, the world is full of error and imprecision, but we can overcome this inherent fallibility through shared effort and dogged communication. As a common tongue now spoken by nations around the world, more widely accepted than any currency or language, the metric system is fundamental to this work. On stage, Phillips reminds us of the founding vision of metrology: 'And so today, we are so much closer to realising that revolutionary dream, *à tous les temps, à tous les peuples*: for all times, for all people.'

As the applause finishes and a vote takes place to certify the kilogram's new definition, I wonder how true this motto really is. The Kibble balance, after all, is hugely expensive and technically challenging to operate. Although this new method can theoretically be used anytime and anywhere to recreate the kilogram from scratch, only a few nations actually have the resources and expertise to do so. Is this really a definition for all people? Certainly not in the way that the revolutionaries meant. Though the premise was flawed even then: the seven-year trek across France to measure the Earth's meridian was useful in part because it was so difficult to repeat. It created a one-time result that could not be challenged by other parties, with the resulting authority of the metre resting, in part, on the time and labour taken to produce it.

The contemporary philosopher Bruno Latour refers to this process as 'blackboxing' – the method by which 'scientific and technical work is made invisible by its own success'. When scientific knowledge becomes a black box, it means we discard the context of its creation. We ignore the human errors, alternative theories, and invisible bodges that are just part and parcel of experimentation, and strip away these uncertainties until only the input and output of the work remain.[44] Everything else is hidden by the black box. This process enhances the authority of science, removing the mess and contingency of research so that results seem singular and objective. It also obscures criticism of the scientific process, hiding controversial or arbitrary decisions. Think about debates about climate change, for example, and how they've stressed the need to limit increases in global temperature to less than 2°C of warming. This is obviously a somewhat arbitrary figure, and the projected outcomes associated with it are far from certain. Things could go a little better or they could go a lot worse. Blackboxing this process creates a clear goal for policymakers and public discussions, but it also gives ammo to conspiracy theorists and 'sceptics' who believe some wonderful truth about the Earth's future is being hidden.

In the world of measurement, this blackboxing of the kilogram's definition might seem retrograde, shutting ordinary women and men out of a field that affects their lives. It's the sort of alienation that motivates anti-metric campaigners and the exact opposite of the medieval *pietre di paragone* that let anyone verify units of measure with their own hands and eyes. But increasing precision in measurement and the subsequent increase in obscurity is a direct result of the fusion of science and industry; a precursor to and product of the benefits of the contemporary world. And this is the bargain we have made for our current comforts. Who cares how units of measurement are defined, as long as the benefits remain? Perhaps one fair way to judge the situation is with Peirce's philosophy of pragmatism, which he once suggested was only a restating of Jesus's wisdom: 'Ye shall know them by their fruits.'[45] In this case, the fruits of modern measurement are the fruits of the modern world, for better or for worse.

In Versailles, the vote in the auditorium is approved, as everyone expected, and the official definition of the kilogram is changed the following year. As the metrologists hoped, nobody who missed the news noticed. Theirs is an invisible discipline, their work hidden from the public view, tucked away at the end of a string of decimal places. Yet on these precarious digits hangs the world of modern measurement, and with it, the frontier of human understanding.

10

THE MANAGED LIFE

---

*Measurement's place in modern society and
in our understanding of ourselves*

> The driving cultural force of that form of life we call 'modern' is the idea, the hope and desire, that we can make the world *controllable*. Yet it is only in encountering the *uncontrollable* that we really experience the world.
>
> —HARTMUT ROSA, *THE UNCONTROLLABILITY OF THE WORLD*[1]

Standardised peanut butter and chlorinated chicken

If anything exemplifies the primacy of measurement in contemporary life, it is the existence of standardised peanut butter. It's the work of the United States' National Institute of Standards and Technology (NIST) and is sold to industry at a price of $927 for three 170-gram jars. The exorbitant cost is not due to rare ingredients or a complex production process, but is instead a result of the rigour with which the contents of each jar have been analysed. This peanut butter has been frozen, heated, evaporated, and saponified, all so it might be measured across multiple dimensions. The result is that when buyers purchase a jar, they can be certain not only of the exact proportions of carbohydrates, proteins, sugars, and fibre in every spoonful, but of the prevalence – down to the milligram – of dozens of different organic molecules and trace elements. This includes familiar names like copper, iron, and magnesium, as well as exotic-sounding substances like docosanoic, eicosanoic, and tetradecanoic acids. Hardly an atom in these jars has avoided scrutiny and, as a result, they contain the most categorically *known* peanut butter in existence. It's also smooth, not crunchy.

The peanut butter belongs to a library of over 1,200 standard reference materials, or SRMs, created by NIST to meet the demands of industry and government. It is a bible of contemporary metrology, each listing testament to the importance of unseen measurement in our lives. Whenever something needs to be verified, certified, or calibrated – whether that is the emissions levels of a new diesel engine or the optical properties of glass destined for high-powered lasers – the SRM catalogue offers the standards against which checks can be made. Most items are mundane: concrete and iron samples for the construction trade; slurried spinach and powdered cocoa for food manufacturers. But others seem like ingredients lifted from God's pantry: ingots of purified elements and pressurised canisters of gases, available in finely graded blends and mixtures. Some are just whimsical, as if they were the creation of an overly zealous bureaucracy determined to standardise even the most unique substances available. Think domestic sludge, whale blubber, and powdered radioactive human lung, available as SRMs #2781, #1945, and #4351.

Each material has a purpose, however. Domestic sludge, for example, is used as a reference by environmental agencies to check pollutant levels in factories. Standardised whale blubber helps scientists track the build-up of chemical contaminants in the ocean (blubber is particularly useful in this regard, as whales are the end point of the marine food chain). Powdered lung, meanwhile, is used as a benchmark for human exposure to radioactive materials. It is both a by-product of, and response to, Cold War fears of nuclear annihilation, and shows how inventive NIST can be in its quest for metrological archetypes. Their samples were created from 70 kilograms of human lung donated by employees of Los Alamos National Laboratory, birthplace of the atomic bomb. Each donor had been exposed to radiation during their life and offered their body to science after death. Like many SRMs, the lungs had to be freeze-dried and pulverised into a fine powder to ensure homogeneity in each sample, which makes for some amusingly straight-faced lab notes. 'Despite the sterilization process that ensured safe handling, it was more than unsettling to occasionally have bits of tissue sprayed on to the laboratory walls and us during the grinding stage, necessitating at least one necktie and lab coat change,' wrote the NIST researcher tasked with the grinding. 'At one point, we even had a red ooze with bits of tissue floating in it making its way down the hall.'[2] Such is the messy, unseen work of making measurements stick.

The purpose of all these materials is to offer clients 'truth in a bottle', says Steve Choquette, director of the agency's Office of Reference Materials (ORM). We're speaking over a video call as the pandemic has locked down NIST headquarters, but even through a computer Steve is an enthusiastic and genial presence. He's keen to discuss the minutiae

In NIST's library of Standard Reference Materials,
you can find reference standards for whale blubber,
powdered radioactive human lung, and peanut butter

of measurement, and knows the demands of the job intimately, having developed optical standards for NIST before moving to the ORM. It's his job, he says, to ensure that customers can have total trust in the agency's measurements. 'We beat these things to death,' he says of the standards, noting that the quantities assigned to each material can ultimately be traced back to the metric system, which, handily, NIST also helps define. If Steve has any questions on the finer points of measurement, he says, he can 'walk across the aisle and talk to the world's experts'.

The SRMs themselves are stored in 25,000 square feet of warehouse space equipped with various grades of freezers, as well as containment areas for radioactive and hazardous material. The catalogue spans multiple categories, explains Steve, but each SRM falls into one of two camps, depending on its use: calibration or validation. Validation means using SRMs to ensure consistency in certain industry tests. Take, for example SRM #1196: the standard cigarette, yours for $218 for a carton of 100. The cigarettes are not used by tobacco companies to compare the quality of their smokes, but by laboratories to test the flammability of fabrics and upholstery. Fires started by stray smoking materials are the leading cause of deaths by fire in the home in the US, says Steve, killing hundreds every year, with nearly half of fires started in the living room and a third in the bedroom.[3] To reduce these hazards, there are various state and industry standards for flame-resistant mattresses, couches, and sheets. But for those tests to be consistent across manufacturers, there needs to be a standardised cigarette to start each fire with. That's SRM #1196. The cigarettes themselves are made by ordinary manufacturers and their contents are no different to those of regular cigarettes. What makes them special is their homogeneity, the predictability of their contents, which is tested and verified by NIST. That's what the agency provides, says Steve: sameness as a service. The results, though, can be incredibly beneficial. 'That product saved a lot of lives once the regulations were enforced,' he says proudly of the cigarettes.

constants of nature, the SRMs are reassuring physical materials, specific in both purpose and definition. Each is created to meet a particular need and derives its utility from certain precise values. Despite their oddity, they fulfil the same basic purpose as the cubit rods of the ancient Egyptians and the bronze weights of Babylonian merchants. They are designed to create uniformity across time, space, and culture; to enable control at a distance and ensure trust between strangers. What has changed since the time of the pyramids is not only the precision of each measure, but the territory they preside over, with present-day standards often encompassing the entire globe.

From the twentieth century onwards there has been a huge, incalculable increase in measurement activity. NIST is only one agency among many, and some 140 countries have their own national standards organisations,[5] with numerous international bodies dedicated to this same work. The International Organization for Standardization, or ISO, for example, regulates some 22,000 standards from its headquarters in Geneva, issuing guidelines for everything from clean rooms used for precision manufacturing and lab work (graded 1 through 9 based on the number of particles occupying each cubic metre of air) to the speed of photographic film (the source of ISO settings on cameras). ISO even publishes ISO 3103, which describes a standard method for making tea. Six pages of instructions include specifications for the type of pot to be used (white porcelain or glazed earthenware), the quantity of dried tea leaves (2 grams per 100 ml of boiling water), and total brewing time (six minutes). The aim is not to create the perfect cuppa, but, as with NIST's domestic sludge and powdered lung, to establish a repeatable baseline for other sorts of test. In this particular case, ISO 3103 is used by food manufacturers who wish to 'examine the organoleptic properties of the infused leaf'.

But why this proliferation of measurement in the first place? According to the few academics who examine metrology in such broad terms, there are a handful of connected explanations. One is the rise of consumer culture, which demands reliable standards for goods in a marketplace flush with choice. When you can't easily trace where the food in the supermarket is from or choose between dozens of different TVs, there needs to be some sort of oversight into their safety and quality. Another points the finger at technological advances, which have both afforded and demanded ever finer degrees of precision. Probably the most important reason, though, is the increase in global trade, which has led to the promulgation of standards as companies seek to reduce complications and costs.

The best illustration of this is the creation of the standardised shipping container, a durable steel box usually 8 feet across, 8 feet 6 inches tall, and 20 or 40 feet long (that's $2.44 \times 2.59 \times 6.10 / 12.19$ metres). These containers were first trialled in the early decades of the twentieth century, before slowly becoming standardised from the 1950s onwards (ISO now handles their specifications). Prior to the introduction of shipping containers, goods had to be packed on to ships by hand, an arduous process that required days of work by crews of longshoremen, adding to costs and slowing the movement of cargo. But being able to pack everything into one-size-fits-all boxes that could be hoisted off ships and on to trucks in minutes significantly lowered the price of moving material around the world, leading to an explosion in shipping. Today, shipping containers are the building blocks of the global economy, the standards that make it economically feasible to manufacture goods in one country, package them in a second, and sell them in a third. For better or worse, it is shipping containers that made fast fashion and the iPhone possible; that created the conditions that allow the world's largest-ever corporations to exist. They helped create the flat world we live in now – not necessarily the one envisioned by Benjamin Constant in the wake of Napoleon's conquests, but the neoliberal economic order described most famously by Thomas Friedman in his 2005 book *The World Is Flat*. According to Friedman and his supporters, this is a world in which barriers of distance, language, and nationality have been removed by technology and corporations. Theirs is the new arrogant eye of power that deplores any obstacle.

NIST itself first got into the standards game because of the demands of business. Specifically, the railroad industry, which expanded rapidly in the US at the end of the nineteenth century, with a corresponding rise in accidents, injuries, and fatalities. Derailments that killed dozens were common, with a death rate of 24 per billion passenger miles in 1896[6] (compared to 0.43 deaths per billion miles for mainline rail in the US today[7]). The reason was simply a lack of standardisation – in protocol, equipment, and material. NIST took charge of the situation in 1901, working with railroad engineers to create the very first SRM: standard metal alloys that would be used as references for the cast-iron wheels of carriages. In combination with other regulatory efforts by the US government, rail fatalities quickly fell. Steve tells me this origin story still reflects how NIST likes to create new standards today, balancing the needs of the public and industry. The agency itself is non-regulatory, meaning it doesn't decide when new standards are needed, but it seems that this suits all involved, keeping NIST's activities out of the political spotlight. 'We've got no axes to grind,' says Steve. 'We just provide the decision-makers with the ability to make decisions.'

Such claims to neutrality should be taken with a pinch of salt, of course: no scientific work is without its biases, and deciding the exact parameters of even a single reference material can have significant ramifications. NIST is currently developing standards that will be used by the US federal government to distinguish between hemp and marijuana, for example, a process that will have significant effects on a number of industries. Both products are derived from the same plant, cannabis, but distinguishing between the two is based solely on the prevalence in the plant of THC, or tetrahydrocannabinol, the chemical that creates the high in humans. A bill in 2018 established that the exact boundary line was 0.3 per cent of THC per total weight. But, as Steve explains, this still leaves open many questions that NIST now has to help settle. 'Do you test that based on dry weight or wet weight? What part of the flower do you test? Are you just looking for THC, because that can oxidise during tests, or do you measure total cannabinoids?' The answer to these questions must draw on culture, economics, and politics. They depend on America's changing attitudes to its failed 'war on drugs'; to demands at the state level to legalise recreational marijuana; and to the subsequent interest of farmers trying to cash in on a new crop. 'This is big business,' as Steve puts it. 'Farmers need to know that they can grow a crop of hemp without having the DEA, the drug enforcement agency, coming in and destroying it. Because they will if it's over the limit.'

Delve just a little into the world of standards and it's clear what a huge, unseen effect they have on all our lives. Scan the news on any given day and you'll find measurement buried in stories about working conditions in warehouses, the approval of new medicine, and trade negotiations. Few recent events in my own lifetime have demonstrated the centrality of standards better than the UK's decision to leave the European Union. Pro-Brexit campaigners argued that doing so would free the country from the over-determined bureaucracy of Brussels, while Remainers responded that leaving the Union would put the UK at the mercy of less rigorous rule-makers. Standards became battle standards, fought over and rallied round. Months were spent arguing over the threat of chlorinated chicken, for example, the end product of certain farming conditions that are acceptable in the US but not in the EU. Whether or not this foodstuff would become the norm in the

UK after Brexit made newspaper headlines, was joked about on panel shows, and debated in Parliament. Like the loss of imperial weights and measures many years before, chlorinated chicken became a surrogate for wider fears about Brexit, framed as another sort of cultural loss and an act of ethical regression. Whether or not it was explicitly articulated, everyone understood that this particular standard – should chicken meat be sold after being given an anti-microbial wash? – was the end product of a vast Rube Goldberg machine of interconnecting economic and political switches. It just so happened that the result of these machinations was one destined to end up on dinner tables around the country.

The metrics of modern life

Despite their abundance, international standards like those mandated by NIST and ISO are mostly invisible presences in our lives. Where measurement does intrude is through bureaucracies of various stripes, particularly education and the workplace. It's in schools where we're first exposed to the harsh lessons of quantification; where we're first sorted by rank and number and told that *these* are the measures by which our future success will be gauged. As with Alfred Binet's IQ tests, there are often beneficial reasons for using such tools, but their simplicity and applicability tends to strip away the thought and care with which they might be applied. Once we leave school and begin work, those same tests reappear in the form of managerial oversight, which squeezes assessment into new and ominous acronyms: KPIs and OKRs. In my own early career as a journalist, the value of my work was judged primarily by a pair of key statistics: the number of articles I wrote per day and the page views these attracted. My peers and I – mostly recent university graduates, badly paid and overworked – were taught to value quantity over quality, learning that what the machine of online journalism demanded was a constant churn of clickable headlines. Adapting to this pressure is something the industry as a whole still struggles with, and I personally had to unlearn many of the lessons taught by these particular metrics.

The underlying principle – that any human endeavour can be usefully reduced to a set of statistics – has become one of the dominant paradigms of the twenty-first century. The historian of capitalism Jerry Z. Muller calls it 'metric fixation', a ubiquitous concept that pervades not only the private sector, but the less quantifiable activities of the state, like healthcare and policing.[8] 'We live in the age of measured accountability, of reward for measured performance, and belief in the virtues of publicising those metrics through "transparency",' writes Muller. And although, as he stresses, measurement itself is not a bad thing, an obsession with measurement above all else will distort, distract, and destroy what we claim to value. 'The problem is not measurement,' writes Muller, 'but excessive measurement and inappropriate measurement – not metrics, but metric fixation.'[9]

The roots of this ideology can be traced back to changes in capitalism beginning in the nineteenth century. This was a period when management in the US particularly was emerging as a profession in its own right, rather than a proficiency learned by industry natives. Between 1870 and 1900, the number of salaried managers in America increased 500 per cent, from 12,501 to 67,706, creating an entirely new type of business structure, which historian Alfred Chandler has identified as 'managerial capitalism'. In contrast to 'personal capitalism', in which those making decisions about a business have a direct stake in its operations, such responsibility is instead outsourced to 'teams, or hierarchies, of salaried managers who had little or no equity ownership in the enterprises they operated'.[10]

A drive to rationalise the work of managers dovetailed with a transformation of industrial production itself. Throughout the nineteenth century, the United States pioneered what became known as the 'American system' of manufacturing – an approach that centred on the virtues of standardisation, precision, and efficiency. Key to this method was the use of interchangeable parts in factories. Previously, the production of consumer goods had been the work of artisans who hand-crafted orders from start to finish. But with the creation of machines that could stamp, cut, and mould components to a high degree of precision, manufacturing could be turned into a series of rote tasks, with lower-skilled workers assembling products piece by piece. As one British engineer who toured US factories in the 1850s noted, wherever the 'aid of machinery' could be 'introduced as a substitute for manual labor, it is universally and willingly resorted to.'[11]

At the turn of the twentieth century, this system was augmented further by two complementary concepts: scientific management and mass production. The latter is best encapsulated by the work of automaker Henry Ford, whose low-priced Model T reshaped not only industrial practice but American culture, helping create a prosperous middle class that defined itself by mass consumption (which would in turn lead to those increasing demands for consumer standards). Ford claimed that his assembly lines, which kept workers static while material moved through their stations on conveyor belts, had been inspired by an aide's visit to a Chicago slaughterhouse. There, the aide observed the opposite process in action: a 'disassembly line' in which a row of butchers took apart pig carcasses, joint by joint, with each individual focusing on a single repetitive task.

This compartmentalisation of labour led to the scientific management movement pioneered by efficiency-obsessed engineer Frederick Winslow Taylor, who advocated a set of working practices now known as Taylorism. Taylor and his followers analysed working practices through 'time and motion studies', which involved

Measurement has become a foundational principle by which we order life and work, as illustrated in Diego Rivera's famous murals of a Ford assembly line

observing labourers and breaking down the flow of their work into constituent parts that could then be standardised. Like the row of butchers that inspired Ford's aide, it was another act of disassembly. The aim, said Taylor, was to 'develop a science to replace the old rule-of-thumb knowledge of the workmen.'[12] Importantly, this also necessitated a transfer of knowledge and power, from the labourers who carried out the work to the managers who oversaw it.

Two of Taylor's most successful disciples, husband and wife Frank and Lillian Gilbreth, even created a new unit of measurement to guide their studies: the therblig, an anagram of their last name, which refers to eighteen fundamental motions that supposedly define all physical activity. 'Suppose a man goes into a bathroom and shaves,' wrote the Gilbreths' children in their autobiographical novel *Cheaper by the Dozen*, which described growing up in the home of efficiency-mad experts. 'We'll assume that his face is all lathered and that he is ready to pick up his razor. He knows where the razor is, but first he must locate it with his eye. That is "search," the first Therblig. His eye finds it and comes to rest – that's "find," the second Therblig. Third comes "select," the process of sliding the razor prior to the fourth Therblig, "grasp." Fifth is "transport loaded," bringing the razor up to his face, and sixth is "position," getting the razor set on his face. There are eleven other Therbligs – the last one is "think!"'[13]

It's the sort of zealous over-measuring that seems as ludicrous as the decimalisation of time, yet it is endemic in society, a tool by which control is exercised not just in the workplace, but in institutions like prisons, armies, and schools. It's in these sorts of measures that French philosopher Michel Foucault anchors his description of the 'disciplinary society', a world in which compliance is enforced by regulating activity using strictly defined norms. In his 1975 text *Discipline and Punish*, he uses changing modes of criminal punishment as an example of this phenomenon, noting that prior to the eighteenth century, obedience was enforced primarily through grisly spectacles like public execution and torture. Those practices have since been replaced by penal systems that instead create 'docile bodies' by controlling individuals' identity and behaviour. Prisoners are given uniforms and numbers, told when and where to eat and sleep, and live in the uncertain knowledge that they are being watched by unseen guards. Eventually, they internalise this authority over their lives and police their own behaviour, says Foucault. Compliance is achieved without overt brutality, but the aim is 'not to punish less, but to punish better'.

This potted history may seem a little abstract, but it's worth remembering the effects such a fixation with measurement can have when brought to bear on certain subjects. One of the most macabre examples of this is perhaps the Vietnam War 'body count', the strategy by which the US military, under the instruction of Secretary of

Defense Robert McNamara, gauged its success in the conflict from a single metric: the number of enemies killed.

This decision to enshrine the body count in this way was a product of both the war's particular conditions and McNamara's ideological bent. The US had been unable to counter the guerrilla tactics of the Viet Cong and so decided to pursue a war of attrition: to inflict unbearable casualties on the communist fighters and drive them north. It was McNamara who believed that progress towards such a goal was best measured by corpses. His own background had straddled the worlds of military and managerial efficiency, with an education at Harvard Business School followed by a job improving the 'efficiency' of US bombing runs in the Second World War. His work included the infamous Operation Meetinghouse during the firebombing of Tokyo, a raid in which an estimated 100,000 civilians were 'scorched, boiled and baked to death' in a single night. (McNamara later admitted, without visible remorse, that this was the work of war criminals.[14]) After the Second World War ended, McNamara moved to the corporate world, spending fifteen years at Ford to rescue the auto-maker from stagnation by cutting costs. Then, in 1961, he was tapped to become Secretary of Defense and overhauled the US military with a mercenary intelligence that could only have come from the world of business. His leadership led to 'the wholesale substitution of civilian mathematical analysis for military expertise', according to historian Edward Luttwak, introducing 'new standards of intellectual discipline and greatly improved bookkeeping methods, but also a trained incapacity to understand the most important aspects of military power, which happen to be nonmeasurable'.[15] The focus instead was on anything that *could* be counted: on dollars spent, firepower amassed, and enemies killed.

Marching under the banner of the body count, US officers in Vietnam crafted a method of warfare that brutalised the country's inhabitants and, in a very different way, the soldiers who carried out this work. Military units were pitted against one another to see who could rack up the most casualties; 'kill boards' were hung in mess halls and printed in army newspapers to keep score; and officers established 'production quotas' for their teams, rewarding soldiers with promotions, crates of beer, and trips to the beach. New tactics like 'free-fire zones' – areas where unidentified individuals were assumed to be enemies unless proven otherwise – were created to market brutality as strategy.

The results were catastrophic. The most benign thing that can be said about the Vietnam body count is that it was fabricated, with soldiers flinging AK-47s on to farmers killed in their crossfire and marking them down as dead Viet Cong in order to meet the quotas set by their superiors. It is more accurate to say that it encouraged war crimes. American troops massacred civilians, children, and babies in the knowledge that they were unlikely to face punishment; they turned unimaginable human suffering into statistics because they understood that this was the nature of their war. As one internal military report noted, 'the pressure to kill indiscriminately, or at least report every Vietnamese casualty as an enemy casualty, would seem to be practically irresistible'.[16] The resulting culture of death was captured best in a song written by soldiers of the 1st Cavalry Division: 'We shoot the sick, the young, the lame, / We do our best to kill and maim, / Because the kills count all the same, / Napalm sticks to kids'.[17]

As with the violence of colonial expansion or the cruelties of the eugenics movement, attributing such behaviour solely to humanity's 'metric fixation' would only be another sort of cowardice. We cannot ignore the role that unthinking trust in number plays in such episodes, but neither should we narrow the scope of our explanation to place blame on a single cause. Savagery of this sort has far deeper roots: in racism and cultural exceptionalism, in centuries of colonial rule and the dehumanisation of the foreign Other. To ignore these factors and abbreviate the story would be to fall into the same trap that awaits those who work solely in numbers, and which, in the end, contributes to such barbarism: it is to oversimplify the world.

Quantified self – the most personal measures

Writing in *The New York Times* in 2010, the technology journalist Gary Wolf heralded our current age of quantification. Using data to make decisions is now the norm in nearly all spheres of life, he wrote. 'A fetish for numbers is the defining trait of the modern manager. Corporate executives facing down hostile shareholders load their pockets full of numbers. So do politicians on the hustings, doctors counseling patients and fans abusing their local sports franchise on talk radio.' Business, politics, and science are all steered by the wisdom of what can be measured, said Wolf, and the reason why is obvious: numbers get results, making problems 'less resonant emotionally but more tractable intellectually'. Only one domain has resisted the lure of quantification, and it is an aspect of society that could sorely benefit from the wisdom of the spreadsheet: 'the cozy confines of personal life'. That, said Wolf, would soon change.[18]

Thanks to new technology – namely, the ability to digitise information, the ubiquity of smartphones, and the proliferation of cheap sensors – humans now have historically unprecedented powers of self-measurement. At the turn of the seventeenth century, Santorio Santorio had to live in a giant pair of scales to produce just a trickle of personal data. Today, we are rewarded with floods of the stuff with minimal effort. We can track our sleep, exercise, diet, and productivity with just a few apps and gadgets. Simply participating in the networks of modern life – owning a smartphone and accessing the internet – turns us into beacons of unseen measurement, emitting quantified data as heedlessly as uranium produces radiation.

For Wolf, the potential of this information is huge. 'We use numbers when we want to tune up a car, analyze a chemical reaction, predict the outcome of an election,' he writes. 'Why not use numbers on ourselves?' His article is the nearest thing to a manifesto for the Quantified Self movement: a loose affiliation of individuals whose pursuit of 'self-knowledge through numbers' shows how far we have internalised the logic of measurement. The movement's origins can be traced back to the 1970s, when enthusiasts cobbled together the clunky ancestors of today's wearable tech. But the idea came to greater public attention after Wolf and fellow journalist Kevin Kelly, founding executive editor of *Wired* magazine, coined the term 'quantified self' in 2007 and founded a non-profit to proselytise their ideas.

Descriptions of the quantified self lend themselves to caricature, creating images of digital Gradgrinds obsessively pursuing the optimised life while their souls wither on the vine. And it's true that many proponents of QS, as it is known, do nothing to dispel this image. They boast about shaving minutes off their day through rigorous self-surveillance, or discovering through sophisticated analyses that – surprise! – good sleep and regular exercise improve their mood. The obfuscation of the obvious has always benefited self-help guides and cults, and the Quantified Self movement sometimes seems to occupy the overlap between these spheres. A typical article from *The Financial Times* in 2011 featured 'self-tracker and bio-engineer' Joe Betts-LaCroix, who tracked, among other things, the daily weight of his wife and two children for three years. His partner, Lisa Betts-LaCroix, told the *FT* that even as she was giving birth, 'instead of holding my hand and supporting me and hugging me, he was sitting in the corner entering the time between my contractions into a spreadsheet.'[19] Such anecdotes give tech critics like Evgeny

Morozov ample evidence to decry the movement as an outreach programme for Silicon Valley 'solutionism' – the belief that error and inefficiency are best overcome by the accumulation and analysis of data. The quantified self is simply 'Taylorism within', says Morozov, and another example of the 'modern narcissistic quest for uniqueness and exceptionalism.'[20]

Proponents of the movement don't deny its introspective qualities, but defend these as a response to the 'imposed generalities of official knowledge.'[21] If quantification has turned the world into generalised rules that do not fit the individual, why not create one's own set of numbers that better capture the truth? They cite anecdotes of self-trackers whose chronic ailments – sleep apnoea, allergies, and migraines – resisted the cures of mainstream medicine but yielded to their pattern-finding prowess. After months and years of diligent self-tracking, these individuals discover some previously hidden mechanism in their life, some food or habit that triggers their affliction, and make the changes necessary to live happily ever after. In this guise, the quantified self seems like an attempt to recapture the personal dimension of measurement; to resist the abstractions of statistics and tailor calculations to fit the contours of our lives. Like the medieval scholar who cured his pain by measuring the dimensions of his body over and over, adherents hope that diligent attention to the self will deliver salvation.

This attention, though, is not always focused, and at times, adherents seem to measure as a reflex. Lurk on the QS forums online and you can find self-trackers who have recorded wonderful trivialities: the cardinal direction they faced every few seconds for three years, for example, the correlation between their anxiety and burp frequency, or the frequency with which friends and relations spring into their mind unbidden. (The last tracker turned these moments into pie charts, finding that they most frequently thought of their parents unprompted.) Such behaviour is not entirely new – just think of Francis Galton's habits of quantification, calculating fidgets per minute and constructing a 'beauty map' of Britain – but technology has afforded greater access. In a different context, some of this activity would look like performance art: a response to, and critique of, the dogma of measurement. But in his 2010 article, Wolf suggests that self-trackers are simply acknowledging the era's prevailing ideology: they live in a world of numbers, so how else are they to understand themselves? Wolf says that a century ago we turned to psychoanalysis to unravel the mysteries of the self, relying on language and a culture of 'prolix, literary humanism'. This, he implies, is not the world we live in today, so why rely on outdated methodologies?

The question he never answers is how the precision of numbers is supposed to match the complexity of language as a tool for self-exploration. One suspects, though, that this is a feature rather than a bug. By limiting the scope of self-investigation to what can be measured, practitioners are assured of finding answers. Supplicants on the therapist's couch, meanwhile, have to return week after week to grapple with the inefficient complexities of language.

It's true, though, that self-trackers are responding to the dominant culture. Their habits – *our* habits, for many of us track our lives, albeit in a less explicit manner – are not just a curious totem of twenty-first-century living. As with the grain measures of the Middle Ages or the land surveys of colonial powers, measurement here is a practice that reveals the superstructures of society. In this case, self-tracking is entangled with the digital control systems of contemporary life; with the vast surveillance machines operated by national intelligence agencies and the workings of tech giants like Facebook and Google. Wolf and Kelly are steeped in the Silicon Valley culture that birthed these firms, and the overlap between their missions and methods is unmistakable. Google is a company with a goal to 'organize the world's information and make it universally accessible and useful' and whose former executive chairman, Eric Schmidt, said that technology is 'not really about hardware and software any more', but about 'the mining and use of this enormous data to make the world a better place'.[22] The only difference between these statements and those of the Quantified Self movement is the scale of their ambition.

In the case of Google's and Facebook's shareholders, making the world a better place primarily means selling adverts. Despite both companies' wild and ambitious side projects, from virtual reality headsets to self-driving cars, their wealth is founded upon something much less utopian and glamorous: targeted advertising, which makes up between 80 and 90 per cent of both firms' annual revenue. The secret sauce in this product is personal data. These companies record your shopping habits, your hobbies, your health problems, your income, who your friends are, what movies you like, whether you've got a degree or not, and countless other trivialities, which are all then used to place advertisements in front of your eyeballs. The academic Shoshana Zuboff describes this business model as 'surveillance capitalism', in which firms offer free or useful services in exchange for personal data that can be used to predict and monetise users' behaviour (the most important prediction being whether or not you will click an advert). You use Google to search for products; Google uses this data to find out about you and then sells access to your screen to other companies.

Google and Facebook started this game, but countless other firms are now playing, and they're not restricting themselves to predicting what ads you'll like. They want to try and predict all sorts of aspects of your behaviour, says Zuboff, with the sectors involved including 'insurance, retail, finance, and an ever-widening range of goods and services companies'.[23] Some of the more exotic examples of this sort of tracking and prediction include a babysitting firm that claims to assess applicants' trustworthiness by scanning their social media and assessing the language they use;[24] a health insurer that promises lower premiums to customers if they wear a fitness tracker and hit a certain number of steps each week;[25] and a recruitment company that uses face-scanning algorithms to gauge candidates' 'enthusiasm' for the job.[26] Such algorithmic decision-making is also being embraced by the state, where it's used in some of the most consequential decisions of modern life, predicting criminal activity, medical outcomes, and even exam results.

Some of these examples may seem benign, or at least constitute a familiar sort of evil. After all, having to smile to get a job is hardly new, and isn't it better that an 'objective' algorithm makes these choices than a fallible human? But this ignores the fact that many of the judgements these systems are making are at best inscrutable and often just plain wrong. Machine learning algorithms, for example, are trained on real-life data and so tend to reflect the biases of society. Facial recognition systems often perform less accurately on non-white faces, and text comprehension programs encode sexist language. In one notorious leaked story from Amazon, the company created an AI program designed to review the CVs of job applicants. Its engineers trained the program to select candidates based on its current employee roster, which skews heavily male. As a result, the software decided that women must be inferior hires and penalised applicants from all-female colleges and those whose résumés included the word 'women's'.[27] The program was scrapped before it was ever deployed, but the story shows the potential for error.

For Zuboff, what is particularly dangerous about these systems is that they form feedback loops that shape our behaviour. Is it right if insurance companies issue gadgets that monitor our health? What about companies that now offer training to job candidates so their interviews appeal specifically to hiring algorithms rather than humans? Other methods of influence are more subtle. One notorious research project published in 2014 by Facebook describes how the company tinkered with the news feeds of roughly 700,000 users. It found that if it showed people posts with 'positive emotional content', they would share positive things themselves. If it showed them posts with 'negative emotional content', their updates, too, would become similarly downbeat. Facebook was initially triumphant, calling its work 'the first experimental evidence for massive-scale emotional contagion via social networks',[28] but it soon downplayed these claims after facing a public outcry about its potential influence on millions of human lives.

These feedback loops aren't just an inconvenience, argues Zuboff, they're a threat to free will. She frames this latter concept as contained in 'the gap between present and future' – the moment where we decide how we will act. Yes, we will get up and go to the gym today. No, we will not buy another pack of cigarettes. If this gap is being foreclosed by unseen algorithmic overlords, does our freedom not suffer? 'I am a distinctive human. I have an indelible crucible of power within me,' says Zuboff. 'I should decide if my face becomes data, my home, my car, my voice becomes data. It should be my choice.'[29] The scale and power of digital companies seems to justify such warnings, but it's hard not to also compare these with fears about statistical fatalism that arose in the nineteenth century. These were triggered by the emergence of Quetelet's average man, a figure who no longer haunts our nightmares. Will the same be true of algorithmic controls in due time? And will this be because the danger was overblown or because we simply accept the constraints they impose? It's clear, at least, that these are not new worries. As Fyodor Dostoevsky's *Underground Man* worries in 1864, soon 'all human actions will then be calculated according to these laws, mathematically, like a table of logarithms'.[30]

For many, the most extreme example of this dynamic can be seen emerging in China, where the country's 'social credit' system aims to shape individual behaviour through digital surveillance. It's an amorphous program that has been running in a series of regional trials since 2009, but the country's ruling communist party aims to unite these efforts eventually. It records a wide array of behaviour, categorising it as either good or bad and rewarding or punishing individuals accordingly. The activity being tracked varies, but the focus is on social responsibility. Users can be marked down for cheating in video games, failing to pay their bills on time, or not cleaning up after their dogs, and rewarded for acts like donating to charity, giving blood, and volunteering. Some programs crunch this data into a single number, an individual's social credit score, which then determines certain privileges and restrictions. People with a good score might be given cheaper access to bike- and car-sharing schemes, free health check-ups, or greater visibility on dating apps, while those with lower scores might be publicly shamed, their children rejected from university, or their travel restricted. In 2019, the National Development and Reform Commission of China reported that it had denied 17.5 million flights and 5.5 million rail journeys to those deemed 'untrustworthy' by the social credit system.[31] It's a rather literal fulfilment of the program's avowed aim: to allow 'the trustworthy to roam everywhere under heaven while making it hard for the discredited to take a single step.'[32]

Reactions to the social credit scheme in the West have largely focused on its dystopian elements: on the apparent ubiquity of the government's surveillance and the arbitrary nature of its judgement. But the picture is more complicated than that. Others note the program's high approval ratings in China, especially among older, wealthier, and better-educated individuals,[33] and how the system provides a sense of trust in a country shaken by the destruction of traditional social structures, a result of both the Communist Revolution and a recent, rapid transition to a more capitalist society. As the historian Theodore Porter notes, quantification and measurement have long been tools that overcome distrust by creating a shared space for communication and minimising the need for personal connections to verify claims.[34] The benefits of the social credit system, essentially scoring trustworthiness based on the government's judgement, fit neatly into this mould. More significant, though, are the clear parallels between China's social credit score and similar systems in the West: of supermarket reward schemes, financial credit scores, and the surveillance capitalism that Zuboff identifies. These, too, dole out privileges and punishments based on our behaviour and seem increasingly unavoidable. The main difference is that in the West, such systems are handled by private institutions rather than government and focus more on our actions in the market than in our community. But perhaps the reason we find China's social credit system so alarming is not that it is unusually dystopic or alien, but because it allows us to clearly see, for the first time, the mechanisms already at work in our own society.

The 10,000-step program

When I think about what measurement means in today's society, how it's used and misused and how we internalise its logic, I often end up thinking about a single figure: 10,000 steps. It's a metric you've probably seen before, which is often cited as an ideal daily target for activity and built into countless tracking apps, gadgets, and fitness programs. Walk 10,000 steps a day, we're told, and health and happiness will be your reward. It's presented with such authority and ubiquity that you'd be forgiven for thinking it was the result of scientific enquiry, the distilled wisdom of numerous tests and trials. But no. Its origins are instead to be found in a marketing campaign by a Japanese company called Yamasa Clock. In 1965, the company was promoting a then novel gadget, a digital pedometer, and needed a snappy name for their new product. They settled on *manpo-kei*, or '10,000-steps meter', the first instance of this metric being used to promote health. But why was this number chosen? Because the *kanji* for 10,000 – and hence the first character in the product's Japanese name, 万歩計 – looks like a figure striding forward with confidence.[35] There was no science to justify 10,000 steps, it seems, just a visual pun.

If the 10,000 steps are an illusion, though, they are a useful one. Research into how many steps a day we *should* pursue offers more finely graded targets, yes. They say 10,000 steps is too low for children but daunting for older adults, and sometimes puts them off exercise altogether.[36] Studies show that for older women, hitting as few as 4,400 steps a day significantly lowers mortality rates, but that no additional benefits accrue after 7,500 daily steps.[37] Despite this, it's abundantly clear that any increased activity is good for us, and that people who do pursue a daily target of 10,000 steps have fewer signs of depression, stress, and anxiety, regardless of whether or not they actually hit this goal.[38] In this light, Quantified Self proponents seem to have a point: if you want to reach people, you need to speak in a language that can be understood. And so, because of the often unwarranted prestige we give numbers, because of their appealing simplicity and because they are trusted, the notion of 10,000 steps has travelled far, probably for the better.

And yet this isn't the whole story. When we take a step back from individual examples like this and look at our systems of quantification and measurement as a whole, problems emerge. Why exactly do we need to follow the precepts of self-tracking at all? Why believe in numbers as the route to a happier life? Self-tracking, for example, often seems like a distraction. For those not suffering from chronic ailments, the lessons of the Quantified Self movement are common sense. 'After many years of self-tracking everything (activity, work, sleep) I've decided it's ~pointless,' tweeted one former adherent, former editor-in-chief of *Wired* magazine Chris Anderson. 'No non-obvious lessons or incentives :('[39] And the question of who gets to indulge in this sort of behaviour is not promising. Demographic data is limited, but reports from Quantified Self conferences suggest attendees are mostly white and middle to upper class.[40] If the quantified self offers freedom, then it's a limited franchise. Meanwhile, some of the most tracked individuals on Earth suffer hugely from it: Amazon warehouse workers and delivery drivers have to account for every second of their working day, and this information is used only to squeeze more labour from them at the cost of their health and well-being. Jeff Bezos flies to space for fun, thanking Amazon employees and customers on his return with the words 'you guys paid for all this',[41] while a pregnant worker in one of his warehouses miscarries after she was pushed against her protests to meet her productivity quota.[42]

The German sociologist Hartmut Rosa suggests that our experience of life in the twenty-first century is increasingly shaped by our desire to control the world; to structure it through empirical observation, rendering it as a series of challenges to overcome. 'Everything that appears to us must be known, mastered, conquered, made useful,' writes Rosa. He says this is most obviously apparent in our constant tracking of our own bodies, but that the same framing increasingly structures how we encounter the world outside ourselves. 'Mountains have to be scaled, tests passed, career ladders climbed, lovers conquered, places visited and photographed ("You have to see it!"), books read, films watched, and so on,' he writes. 'More and more, for the average late modern subject in the "developed" western world, everyday life revolves around and amounts to nothing more than tackling an ever-growing to-do list.'[43]

This mindset, says Rosa, is the result of three centuries of cultural, economic, and scientific development, but these trends have become 'newly radicalised' in recent years thanks to digitalisation and the ferocity of unbridled capitalist competition.[44] The history of measurement tracks much of these developments, for not only is it a tool that has been embraced to better understand and control reality, but it now mediates much of our experience of the world, and, crucially, our experience of ourselves. As we measure more and more, we encounter the limits of this practice and wrestle with its disquieting effects on our lives. As noted by Rosa, these problems have been described in many forms by many thinkers over the centuries. For Karl Marx, it takes the form of alienation in our working lives, as we are separated from the products of our labour; for Max Weber, it is understood as the disenchantment of the world, in which the rationalisation of nature removes its magic and its meaning; and for Hannah Arendt, it is the distance created by science and technology that replaces the closeness of human intersubjectivity, of a world previously experienced communally alongside fellow human beings. Rosa himself identifies this unease as 'points of aggression' – moments when our 'efforts and desire to make things and events predictable, manageable, and controllable' are thwarted by 'an intuition or longing to simply let "life" happen'.[45] They are the moments when the learned practices of measurement collide with something deeper and unquantifiable in ourselves.

At the beginning of this book, I said that my interest in the subject was piqued by a simple curiosity about the origin of certain units of measurement. Why is a kilogram a kilogram, I asked; why an inch an inch? I understand these questions more fully now, for if measurement is the mode by which we interact with the world, then it makes sense to ask where these systems come from and if there is any logic to them. The answer I've found is that there isn't any – not really. Or rather, there is logic, but it is, like the 10,000 steps, as much the product of accident and happenstance as careful deliberation. The metre is a metre because hundreds of years ago, during the French Revolution, certain intellectuals thought that defining a unit of length by measuring the planet we live on was the most rational course of action. And the metre is also a metre because mistakes made during that expedition have been enshrined in our system of measurement ever since. In other words: it's the way it is because we say it is.

This is comforting, I think, or it can be. It is a reminder that the systems of the world, those frameworks of order that seem inviolable because of their deep roots in tradition and authority, are as changeable as anything else in life. Like every grand plan dreamed up by humanity, every scheme and method of control and organisation, they can be questioned and altered. They have been shaped by the complexities of the world and the unpredictability of human lives, and that makes them fallible, just as we are. If they don't work – if they don't measure up – then they too can be remade.

EPILOGUE

The measures in the head

As you might guess from the fact that I've written a book about it, I'm something of a measurer myself, though I like to think I'm no caricature. I don't religiously count steps or calories, I'm not fastidious in my appearance, and though I'm told I used to obsessively line up my toys by order of size when I was a child, it's a habit I've grown out of since. No, I don't think I measure in any obvious way, but measurement is nevertheless a central part of my life, a structuring force that I use, like so many of us, to make sense of the world.

My own personal habits revolve around the prosaic measures of work, fitness, and productivity. I am a list writer and note keeper, scattering my aspirations and to-dos across numerous apps and notebooks. I maintain daily checklists, rewriting them every couple of months when whatever improving habits I'm attempting to adopt fall out of favour. And I exercise by the number, counting reps and weights in the gym, measuring out times and distance on my runs. There's a sense of achievement to it all, which is certainly not shallow (not always anyway). The rows and columns of my checklists help me corral time, herding stray hours into their pens like lost sheep, while calculations in the gym offer the relief of progress that can be measured. These habits give me structure when I feel lost and purpose when I stagnate. In such moments, it is a comfort to turn your life into statistics; to define your accomplishment on paper, or to render fat and muscle as numbers that can be changed with the sweaty application of effort. Such strategies are normal now, I think. Hartmut Rosa is right in his diagnosis of our current condition: in today's world, we measure reflexively and internally. As the curators of the Archives nationales in Paris told me of the schoolchildren puzzled by the existence of a physical metre standard: 'They don't understand why it has to be an object because it's already in their heads.'

The question, though, of what should be measured is harder to answer. Along with my self-flagellating but socially acceptable monitoring of productivity, I have other, less structured forms of measure that are a little more personal, small tests of self-discipline that I developed when young, and which have survived into my adult years. For example, if I see a word I don't know and fail to look up its meaning, it means I'm uninterested in the world; if I eat an apple but not the core, it means I lack resolve. Such tests, I think, are perfectly normal, and we all develop such internal metrics – some hidden and some explicit – to reflect on our own behaviour. They are the spinning regulators of the soul that keep the engines of our ego chuffing along within reasonable limits. The problem is when they become obsessions in their own right.

The silliest of my internal tests was also the most extravagant: a little ritual constructed around Beethoven's Ninth Symphony. I first heard the piece when I was about fifteen and was instantly smitten. I found the fourth movement in particular, the 'Ode to Joy', almost too powerful to bear. As I listened to its theme approaching, triangles tinkling and bassoons parping like a circus on the march, I would become utterly transfixed – ecstatic, even, as the music stretched every nerve of my body on a tuning peg. I became determined to capture this feeling and amplify it somehow, and decided that I would only listen to 'Ode to Joy' when I was magnificently happy, when something positive had happened in my life or I'd achieved something noteworthy. It would be both test and reward, allowing me to mark the high points of my life but also increase my enjoyment of the music by associating it with only good feelings. A friend later pointed out to me that this was essentially a reverse Ludovico Technique, the aversion therapy from Anthony Burgess's novel *A Clockwork Orange*, which also relies on 'Ode to Joy' to underscore its lessons. I don't remember copying Burgess but, as I say, I was fifteen, and teenagers' souls breed such melodrama of their own accord.

The early results of my experiment were positive. I listened to 'Ode to Joy' after obvious milestones – exam results, university acceptance, and so on – revelling in the music each time. As I got older, though, and moved out of the structured world of school and university, these milestones became more infrequent. My expectations for my own happiness became more rarefied and, eventually, distant from me. In my early twenties I fell into a long period of depression, a block of the sort of bad feelings I'd flirted with as a teenager but now felt committed to for good. Nothing in my life was worth celebrating, I thought, nothing I did was good enough or deserving of happiness. As a result, I did not listen to 'Ode to Joy' for five years.

My commitment to the test was absolute. If I heard those familiar opening notes, that heartbreakingly confident rise and fall, I would remove myself immediately. I muted videos, changed TV channels, and left the cinema when it appeared in a film. Once, I even had to jog away from a band playing it in the street. Listening to 'Ode to Joy', I told myself, would have been cheating: an unearned pleasure that I did not deserve. The music was too good, and I was too bad. As the opening soloist sings: '*O Freunde, nicht diese Töne!*' – 'Oh friends, not these sounds!'

Over time, my depression eased. I forgot most of my little tests and developed routines that were more generous in form and frequency. One day, I had some particularly good news: something I'd thought often about as a teenager and that I'd bizarrely, unexpectedly, achieved. I was by myself and decided I needed to mark the moment with a little celebration. On my lunch break I went for a walk to a nearby park, and there, sitting on a bench in the sunshine, I put on my headphones and cued up 'Ode to Joy' for the first time in many years. Yet as the chorus began its familiar strains – *Freude, schöner Götterfunken, Tochter aus Elysium* – I felt . . . nothing. No flicker of electricity, no supernova of emotion, just a hollow in my head as I stared at a squirrel skittering about in the sunshine. Something about the test had broken my capacity to enjoy the music. I'd overburdened it with meaning, piled so much weight on to each note that the whole edifice had collapsed, and with it my enjoyment. I took out my earphones and left.

I think that many of us have this sort of relationship with measurement in our lives, particularly self-measurement. We erect scaffolds of to-do lists and deadlines that are equal parts obligation and aspiration, and construct within their frame the person we want to be. We're encouraged to do so perpetually; tips and guidance on how to be more productive, to achieve more, permeate culture. They fill our magazine pages and social media feeds, promising that this or that new method will be the key to greater productivity and personal fulfilment. This isn't a new phenomenon by any means, but it is ferocious in its current onslaught. Increasingly, our ability to manage our time productively is seen not just as an advantage but as a virtue – judgement on our moral worth. The ancient Egyptians may have been the first to suggest that you could weigh the value of a soul, but thousands of years later such reckoning is ubiquitous.

History shows that the borders of measurement's domain are not fixed. They've expanded as scientists have learned the rewards of observation and flexed to accommodate folklore and mysticism. And while it's no longer common to attribute miracles to measurement, as with the *mensura Christi* and saints' tales of the Middle Ages, there is still a residue of that same magical thinking in how we treat measures today. We have a tendency to venerate numbers for their supposed objectivity, to believe that all of life's problems are soluble with statistics. But sometimes, as we measure something's place in the world, the marks become stronger and the thing itself fades into the background. The plan subsumes the goal and we lose sight of what we wanted in the first place. In a world suffused with measurement, *built* on measurement, we need to remember what purposes it serves and whom it should ultimately benefit

For what it's worth, as I write these words, the last in the book, I'm listening to 'Ode to Joy' again, burning with pleasure.

ACKNOWLEDGEMENTS

While many books are certainly solo works of endeavour – lonely climbs over challenging terrain – I have been reminded while writing these acknowledgements just how much I owe others in helping me to complete this work. *Beyond Measure* is undoubtedly a synthesis, from start to finish; the topic at hand is too sprawling and unwieldy for my academic contribution to be anything but a drawing together of the research and insights of others. But it is also a synthesis of another kind: a product of the patience, wisdom, and kindness of those whom I have worked with and those I love. To all of you – thank you.

First, thanks are due to my wonderful agents Sophie Scard and Kat Aitken, and to Sophie in particular for the sage and indispensable advice that helped sharpen my proposal at its earliest stages. I'm glad we rolled the dice. To my editors at Faber, Laura Hassan and Emmie Francis, I owe the rest of the construction process, from large-scale demolitions to sentence-level repointing. It has been incredible to work with such attentive, incisive, and generous readers, and you have borne my ceaseless fretting and prattling with what can only be described as magisterial grace. My sincere thanks go also to my copyeditor Jenni Davis and my typesetter and proofreader Ian Bahrami, who both made many necessary interventions and saved me from more embarrassment than I would have been able to process. Thanks to Beth Dufour for her tireless work chasing permissions, to Jonny Pelham for his stunning cover design (it seems unfair that it's my name that goes on it, not yours), and to the wonderful team at Faber who have guided this book through proofing, printing, publicity, and into the unknown: Joanna Harwood, Hannah Turner, John Grindrod, Mo Hafeez, and Josh Smith.

I also owe thanks to my colleagues at *The Verge*, who have supported this endeavour directly and indirectly by allowing me time to research and write, sending me on that first reporting trip to cover the redefinition of the kilogram, and helping me learn this trade in the first place. Thanks to Nilay Patel and Dieter Bohn for putting together such a fine crew; to Thomas Ricker for his support and guidance, over the years and the waters; to T. C. Sottek for many kindnesses shown, both personal and bureaucratic; and to Liz Lopatto for, well, educating me in how to be a journalist. (I'm sorry it didn't all stick.)

Beyond Measure draws on many sources, and to the experts and academics I spoke to while researching, writing, and editing – particularly those who looked over chapter drafts and supplied thoughts and corrections – I cannot offer more gratitude. Any insights gleaned herein are the work of better minds, and any mistakes entirely my own. I would like to thank (in order of the material addressed as it appears in the book) Salima Ikram for introducing me to Cairo and the magnificent Roda nilometer; Denise Schmandt-Besserat for her lucid explanation of the origin of writing in clay tokens; Nicola Ialongo for his thoughts on the uncanny synchronisation of Bronze Age weights; Eleanor Janega for her wisdom on all things medieval and her tireless crusade against the misnomer that is 'the dark ages'; Mark Thakkar for his enthusiasm and erudition on the subject of the Oxford Calculators, relayed on a sopping wet afternoon outside the British Museum; Edith Sylla for her time discussing those same wily Oxonians; Tom Ainsworth for his Socratic wisdom correcting my misunderstandings of the ancient Greeks (and Harriet Ainsworth for the introduction to Tom); Emanuele Lugli for his wonderful work on the *pietre di paragone* and his conceptual brilliance outlining the invention of sameness; Anna-Zara Lindbom for her time at the Gustavianum and her knowledge of the sweep of Swedish metrology; Elizabeth Neswald for her scrupulous understanding of the history of thermodynamics and her insights into contemporary cultural reactions; Hasok Chang for his work on the creation of temperature, and for kindly easing me into the world of metrology as my first interviewee for this book; Michael Trott for his time discussing the metric revolution, and whose research on the same is fascinating and informative; Sabine Mealeau and Stéphanie Marque-Maillet of the Archives nationales for their unlocking of locks and introduction to the kilogram and meter of the Archives; Julia Lewandoski for crucial nudges that helped deepen my understanding of the colonisation of the Americas (and for pointing me towards sources that weren't fifty years out of date); Neil Lawrence for his instruction on matters statistical, which I hope I condensed and conveyed without loss of accuracy; Tony Bennett for showing me in person just how ARM gets the job done and for his honest and engaging company over pints; Martin Milton and Richard Davis for welcoming me to BIPM headquarters in Paris and showing me round the artefacts and archives; Stephan Schlamminger for his explication of modern metrology and his scrutiny of my work down to the Planck length; and Steve Choquette of NIST for his guidance through the bizarre and wonderful catalogue of standard reference materials. I'm only sorry I was unable to visit in person and sample that peanut butter myself.

As a short addendum on this theme, while I realise that thanking the authors of the books I have read would be silly, *nevertheless* I would like to do so – at least in spirit. More concretely, I would like to single out a few writers whose work in metrology showed me the possibilities of the subject. Firstly, Robert P. Crease, whose book *World in the Balance* was my introduction to its marvels; secondly, Ken Alder, whose academic rigour and narrative skill in *The Measure of All Things* secures the invention of the metric system as an overlooked event of world-historical significance; and thirdly, Simon Schaffer, whose synthesis of history, culture, science, and philosophy in various writings (a number of which concern metrology) provided a glow of inspiration when my own intellectual curiosity dimmed. In my experience, each page of book-writing requires several dozen of book-reading to supply the necessary matter for growth, and I only hope to have added to this compost. On that note, thanks are due to the staff of the British Library, who maintained the building as a safe port during the pandemic, and who were ever helpful while wrangling microfilm and battered journals.

Lastly, I thank with fervour my early readers, and my friends and family, whose encouragement and support have been irreplaceable. Intellectual inspiration is a fine thing, but actually being told your writing is worth the effort at all is indispensable. Thanks are due, then, to Robert Macfarlane, who read some of the book's first chapters and gave typically generous advice and encouragement at a time when it was sorely needed; to Gavin Jackson, who helped put backbone into many jelly-like chapters and who has always been ready with a sympathetic pint and cigarette; to Rachel Dobbs, who read enough drafts of the Epilogue to constitute cruel and unusual punishment, yet still managed to offer sane advice, both linguistic and emotional; and to my mother, Bridget, who read every word of the thing, and whose ability to shape language with wisdom and care has not diminished since I was a teenager struggling with school essays. Thank you also to my friends, to everyone who has ever listened to me ramble about abstraction versus particularity or complain about writing, or offered me a shoulder to cry on (literal or metaphorical).

It truly means so much. I must single out Tom Crooke for his steadfast friendship through so many verses of life; Ally Daniel, Oliver Ilott, and John Ledger for their particular support and kindness; the boys of York, Matthew Smurthwaite and Fran Lloyd-Jones; and all the attendees of a certain dinner party (I really did think of it often when writing). To Lucie Elven a great deal is owed, but suffice to say without your love and encouragement this book would never have existed: thank you. Last, but not least, I wish to thank my wonderful family for everything they have given me over the years – love, support, and all the rest that cannot be measured: Will and Beth and Alf; Rose and Claudia and George; and my incredible parents, Bridget and John.

NOTES

INTRODUCTION

1 A. S. Brooks and C. C. Smith, 'Ishango Revisited: New Age Determinations and Cultural Interpretations', *The African Archaeological Review*, 5(1), 1987, pp. 65–78. Caleb Everett, *Numbers and the Making of Us: Counting and the Course of Human Cultures* (Cambridge, MA, and London: Harvard University Press, 2017), p. 36.

2 Prentice Starkey and Robert G. Cooper Jr., 'The Development of Subitizing in Young Children', *British Journal of Developmental Psychology*, 13(4), November 1995, pp. 399–420.

3 K. Cooperrider and D. Genter, 'The Career of Measurement', *Cognition*, 191, 2019.

4 Jean Piaget, Bärbel Inhelder, and Alina Szeminska, *The Child's Conception of Geometry* (London: Routledge, 1960, digital edn, 2013), pp. 88–215.

5 Sana Inoue and Tetsuro Matsuzawa, 'Working Memory of Numerals in Chimpanzees', *Current Biology*, 17(23), 2007, pp. R1004–5.

6 D. Biro and T. Matsuzawa, 'Chimpanzee Numerical Competence: Cardinal and Ordinal Skills', in *Primate Origins of Human Cognition and Behavior* (Japan: Springer, 2001, 2008), pp. 199–225.

7 Sir William Thomson, *Popular Lectures and Addresses*, vol. 1 (New York: Macmillan & Co., 1889); 'Electrical Units of Measurement', a lecture delivered at the Institution of Civil Engineers, 3 May 1883 (London and Bungay: Richard Clay & Sons Limited), p. 73.

8 Francesca Rochberg, *The Heavenly Writing: Divination, Horoscopy, and Astronomy in Mesopotamian Culture* (Cambridge: University of Cambridge Press, 2004), p. 260.

9 David C. Lindberg, *The Beginnings of Western Science*, second edn (Chicago: University of Chicago Press, 2008), pp. 12–20.

10 Victor E. Thoren, *The Lord of Uraniborg: A Biography of Tycho Brahe* (Cambridge: Cambridge University Press, 1990), p. 23 (the duel), p. 39 (wealth), p. 345 (the elk).

11 John E. Clark, 'Aztec Dimensions of Holiness', in Iain Morley and Colin Renfrew (eds), *The Archaeology of Measurement* (Cambridge: Cambridge University Press, 2010), pp. 150–69.

12 Cooperrider and Genter, p. 3.

13 Jan Gyllenbok, *Encyclopaedia of Historical Metrology, Weights, and Measures*, vol. 2 (Basel: Birkhäuser, 2018), p. 1,076. About 5 km in distance, though the term *peninkulma* now means a 'mil', or 10 km. Variously spelled *peninkuorma, peninkuulema* or *peninkuuluma*.

14 Eric Cross, *The Tailor and Ansty* (Cork: Mercier Press, 1942 (1999)), p. 115.

15 As historian and science writer Robert P. Crease has calculated it, and from whom I take this example. From: Robert P. Crease, *World in the Balance: The Historic Quest for an Absolute System of Measurement* (New York and London: W. W. Norton & Company, 2011), p. 24.

16 Cross, p. 115.

17 Eric Hobsbawm, *The Age of Extremes: 1914–1991* (London: Michael Joseph, 1994 (Abacus, 1995)), p. 57.

18 John Thomas Smith, 'Biographical Sketch of Blake', in Arthur Symons, *William Blake* (New York: E. P. Dutton and Company, 1907), p. 379.

19 Anthony Blunt, 'Blake's "Ancient of Days": The Symbolism of the Compasses', *Journal of the Warburg Institute*, 2(1), July 1938, p. 57.

20 Max Horkheimer and Theodor W. Adorno, *Dialectic of Enlightenment* (Stanford: Stanford University Press, 2002), p. 182.

21 Horkheimer and Adorno, pp. 4–5.

22 Jonathan Swift, 'Cadenus and Vanessa' (1726), in *Miscellanies*, vol. 4 (London: printed for Benjamin Motte and Charles Bathurst, 1736), p. 121.

CHAPTER 1: THE KINDLING OF CIVILISATION

1 Charles Dudley Warner et al. (comp.), 'Book of the Dead', spell 125 ('The Negative Confession'), *The Library of the World's Best Literature, An Anthology in Thirty Volumes*, Francis Llewellyn Griffith (trans.) (New York: Warner Library, 1917), p. 5,320. https://www.bartleby.com/library/.
Alternatively: 'Homage to thee, O Great God, Lord of Maati, I have come to thee, O my Lord, that I may behold thy beneficence [...] I have come to thee, and I have brought [truth] to thee [...] I have not diminished from the bushel. I did not take from or add to the acre-measure. I did not encroach on the fields [of others]. I have not added to the weights of the scales. I have not misread the pointer of the scales.' E. A. Wallis Budge and A. M. Epiphanius Wilson, *The Ancient Egyptian Book of the Dead* (New York: Wellfleet Press, 2016), pp. 21–6.

Alternatively: 'I swear unto the Soul of Isis, that I have not altered the Sacred Cubit of my fathers.' Found in Charles A. L. Totten, *An Important Question in Metrology Based upon Recent and Original Discoveries: A Challenge to 'The Metric System,' and an Earnest Word with the English-Speaking Peoples on their Ancient Weights and Measures* (London: John Wiley & Sons, 1884).

Alternatively: 'I have not shortened the cubit.' Charles Piazzi Smyth, *Life and Work at the Great Pyramid During the Months of January, February, March, and April, A.D. 1865*, vol. III (Edinburgh: Edmonston & Douglas, 1867), p. 430.

Or: 'I have not reduced the measuring-vessel, I have not reduced the measuring cord / I have not encroached on the fields; I have not added to the pan of the scales.' *Book of the Dead*, chapter 125A. https://www.ucl. ac.uk/museums-static/digitalegypt/literature/religious/bd125a.html.

2 Manfred Lurker, *An Illustrated Dictionary of the Gods and Symbols of Ancient Egypt* (London: Thames & Hudson, 1974), p. 57.

3 Rosa Lyster, 'Along the Water', *London Review of Books*, 43(9), 6 May 2021.

4 Sources: Helaine Selin (ed.), *Encyclopaedia of the History of Science, Technology, and Medicine in Non-Western Cultures*, second edn (Berlin, New York: Springer, 2008), pp. 1,751–60; Horst Jaritz, 'The Nilometers of Ancient Egypt – Two Different Types of Nile Gauges', *16th International Congress on Irrigation and Drainage, Cairo, Egypt, 1996*, pp. 1–20; William Popper, *The Cairo Nilometer, Studies in Ibn Taghrî Birdî's Chronicles of Egypt: 1* (Berkeley and Los Angeles: University of California Press, 1951).

5 Zaraza Friedman, 'Nilometer', in Helaine Selin (ed.), *Encyclopaedia of the History of Science, Technology, and Medicine in Non-Western Cultures*, third edn (Dordrecht: Springer Science+Business Media, 2014), p. 4.

6 John Bostock and H. T. Riley, *The Natural History of Pliny* (London: Taylor and Francis, 1855), Chapter 10 – The River Nile.

7 Friedman, 'Nilometer', p. 5.

8 Denise Schmandt-Besserat, 'The Earliest Precursor of Writing', in William S. Y. Wang (ed.), *The Emergence of Language: Development and Evolution* (New York: W. H. Freeman and Company, 1991), pp. 31–45.

9 Denise Schmandt-Besserat, 'The Token System of the Ancient Near East: Its Role in Counting, Writing, the Economy and Cognition', in Morley and Renfrew, *The Archaeology of Measurement*, p. 29.

10 Interview with the author, 27 August 2019.

11 Denise Schmandt-Besserat, *How Writing Came About* (Austin, TX: University of Texas Press, 1996).

12 Samuel Noah Kramer, *From the Tablets of Sumer: Twenty-Five Firsts in Man's Recorded History* (Indian Hills, CO: Falcon's Wing Press, 1956), p. xix.

13 Christopher J. Lucas, 'The Scribal Tablet-House in Ancient Mesopotamia', *History of Education Quarterly*, 19(3), Autumn 1979, p. 305.

14 Jack Goody, *The Domestication of the Savage Mind* (Cambridge: Cambridge University Press, 1977), p. 81.

15 Goody, p. 94.

16 Alan H. Gardiner, *Ancient Egyptian Onomastica*, vol. 1 (Oxford: Oxford University Press, 1947).

17 Gardiner, pp. 24–5.

18 Goody, p. 102.

19 Jorge Luis Borges, 'John Wilkins' Analytical Language', in Eliot Weinberger (ed.), *Selected Non-Fictions*, Esther Allen, Suzanne Jill Levine, and Eliot Weinberger (trans.) (New York: Penguin, Viking, 1999), pp. 229–32. The essay was originally published as 'El idioma analítico de John Wilkins', *La Nación*, Argentina, 8 February 1942, and republished in *Otras inquisiciones*.

20 Michel Foucault, *The Order of Things* (London and New York: Routledge, 2005 (1966)), p. xxi.

21 Though circadian rhythms are *not* ubiquitous. See Guy Bloch, Brian M. Barnes, Menno P. Gerkema, and Barbara Helm, 'Animal Activity Around the Clock with No Overt Circadian Rhythms: Patterns, Mechanisms and Adaptive Value', *Proceedings of the Royal Society*, 280(1765), 22 August 2013.

22 Incidentally, the ISS is kept on Coordinated Universal Time, or UTC, as a convenient halfway point between the main mission control centres in Houston and Moscow. Ally, '20 Questions for 20 Years: Happy Birthday International Space Station', European Space Agency, 21 November 2018. https://blogs.esa.int/alexander-gerst/2018/11/21/spacestationfaqs/

23 Iain Morley, 'Conceptualising Quantification Before Settlement: Activities and Issues Underlying the Conception and Use of Measurement', *The Archaeology of Measurement*, p. 17.

24 Munya Andrews, *The Seven Sisters of the Pleiades: Stories from Around the World* (Australia: Spinifex, 2004). Clare Oxby, 'A Review of African Ethno-Astronomy: With Particular Reference to Saharan Livestock-Keepers', *La Ricerca Folklorica*, 40, October 1999, pp. 55–64. doi:10.2307/1479763

25 Hesiod, *Theogony; Works and Days; Shield*, Apostolos N. Athanassakis (trans.) (Baltimore and London: Johns Hopkins University Press, 1983 (2004)). *Works and Days*, 615, p. 80.

26 Mark Edward Lewis, 'Evolution of the Calendar in Shang China', in Morley and Renfrew, *The Archaeology of Measurement*, pp. 195–202.

27 E. C. Krupp, *Echoes of the Ancient Skies: The Astronomy of Lost Civilizations* (New York: Dover Publications, 2003 (1983)), p. 205.

28 'What Did Mayans Think Would Happen in 2012?' Interview with Dr Mark Van Stone, KPBS, 6 September 2010. https://www.kpbs.org/news/2010/sep/06/what-did-mayans-think-would-happen-2012/.

29 A lack of leap years in this calendar meant that it slowly crept out of sync with the solar year, earning it the nickname of the *annus vagus*, or wandering year.

30 Alan B. Lloyd, *Ancient Egypt: State and Society* (Oxford: Oxford University Press, 2014), p. 322.

31 Heidi Jauhiainen, 'Do Not Celebrate Your Feast Without Your Neighbors: A Study of References to Feasts and Festivals in Non-Literary Documents from Ramesside Period Deir el-Medina' (PDF), *Publications of the Institute for Asian and African Studies*, No. 10 (Helsinki: University of Helsinki, 2009); charms mentioned on p. 198.

32 Lynn V. Foster, *Handbook to Life in the Ancient Maya World* (New York: Facts on File, 2002), p. 253.

33 This particular example, though not its explication in this form, comes from David Brown, 'The Measurement of Time and Distance in the Heavens Above Mesopotamia, with Brief Reference Made to Other Ancient Astral Sciences', in Morley and Renfrew, *The Archaeology of Measurement*.

34 *The Epic of Gilgamesh*, an English version, with an introduction by N. K. Sandars (Harmondsworth and New York: Penguin, 1977), pp. 209–10.

35 Crease, pp. 18–25.

36 William Rossi, *Professional Shoe Fitting*. Though, contra to some myths, Edward II did not introduce this standardisation, and the first recording of the third-of-an-inch increment Rossi finds is from 1856.

37 Not to be confused with the measure of gold purity that shares the same root, spelled karat in the US and carat in the UK.

38 Diamond Jenness, *The Ojibwa Indians of Parry Island, Their Social and Religious Life* (Ottowa: National Museum of Canada, 1935; Bulletin no. 78, Anthropological Series, no. 17), pp. 11–12.

39 Theodore M. Porter, *Trust in Numbers: The Pursuit of Objectivity in Science and Public Life* (Princeton, NJ: Princeton University Press, 1995), p. ix.

40 David N. Keightley, 'A Measure of Man in Early China: In Search of the Neolithic Inch', *Chinese Science*, 12 (1995), pp. 18–40; p. 26 mentions

gendered units, e.g. 'Moving back ten *fen* from a man's hand' and 'The hand of an average woman is eight *cun*-inches long; one calls this a *zhi*-foot'; Crease mentions the same, p. 37.

41 R. Pankhurst, 'A Preliminary History of Ethiopian Measures, Weights, and Values', *Journal of Ethiopian Studies*, 7(1), January 1969, p. 36. Numerous anecdotes are cited, e.g. 'Petros Wontamo says that in the Misghida area of Kambata, "Every village has its own man on whom the villagers depend for buying cloth ... Such a person is quite a standard of measurement. He is not paid for his assistance, except by a word of thanks."'

42 Herbert Arthur Klein, *The Science of Measurement: A Historical Survey* (New York: Dover Publications, 1974), p. 44. The actual length is 'mesouret at the rut of the nayll', and the thumb itself was probably pressed flat (and thus squashed out a little) before measurement.

43 Mark Lehner, 'Labor and the Pyramids: The Heit el-Ghurab "Workers Town" at Giza', in Piotr Steinkeller and Michael Hudson (eds), *Labor in the Ancient World*, vol. V (ISLET-Verlag, 2015), pp. 397–522.

44 Lewis Mumford, *The Myth of the Machine: Technics and Human Development* (New York: Harcourt Brace Jovanovich Inc., 1967), p. 11.

45 Mumford, p. 12.

46 Mumford, p. 168.

47 Raffaella Bianucci et al., 'Shedding New Light on the 18th Dynasty Mummies of the Royal Architect Kha and His Spouse Merit', *PLoS One*, 10 (7), 2015.

48 Naoko Nishimoto, 'The Folding Cubit Rod of Kha in Museo Egizio di Torino, S.8391', in Gloria Rosati and Maria Cristina Guidotti (eds), *Proceedings of the Eleventh International Congress of Egyptologists, Florence, Italy, 23–30 August 2015* (Oxford: Archaeopress Publishing Ltd, 2017), pp. 450–6.

CHAPTER 2: MEASURE AND THE SOCIAL ORDER

1 For more detail, see Guitty Azarpay, 'A Photogrammetric Study of Three Gudea Statues', *Journal of the American Oriental Society*, 110(4), October–December 1990, pp. 660–5.

2 Arvid S. Kapelrud, 'Temple Building, a Task for Gods and Kings', *Orientalia*, Nova Series, 32(1), 1963, pp. 56–62.

3 John M. Lundquist, 'The Legitimizing Role of the Temple in the Origin of the State', Society of Biblical Literature Seminar Papers 21, 1982, pp. 271–97.

4 Nicola Ialongo, Raphael Hermann, and Lorenz Rahmstorf, 'Bronze Age Weight Systems as a Measure of Market Integration in Western Eurasia', *PNAS*, 6 July 2021.

5 Emanuele Lugli, *The Making of Measure and the Promise of Sameness* (Chicago: University of Chicago Press, 2019), p. 91.

6 L. W. King (trans.), Yale Law School, *The Avalon Project, Documents in Law, History and Diplomacy: The Code of Hammurabi*. Laws 108 and 155.

7 Lugli, p. 142.

8 'Magna Carta 1215', 6.13, *Medieval Worlds: A Sourcebook*, Roberta Anderson and Dominic Aidan Bellenger (eds) (London: Routledge, 2003), 35, p. 156.

9 Mishneh Torah, Laws of Theft 7:12. Maimonides.

10 Howard L. Goodman, *Xun Xu and the Politics of Precision in Third-Century AD China* (Leiden, Netherlands: Brill, 2010), p. 197.

11 Goodman, p. 209. 'Absolute pitch for early China, and even through the whole imperial period, was laden with cosmic significance. The foundation pitches of a corrected set of lü [. . .] and the notes played in court performances were part of the court's religious system.'

12 Goodman, p. 159.

13 Quoted in Goodman, p. 205.

14 Witold Kula, *Measures and Men*, R. Szreter (trans.) (Princeton, NJ: Princeton University Press, 1986), p. 127.

15 Kula, p. 33.

16 Kula, p. 30.

17 M. Luzzati, 'Note di metrologia Pisana', in *Bollettino Storico Pisano*, XXXI–XXXII, 1962–3, pp. 191–220: pp. 208–9, 219–20.

18 Kula, pp. 43–70.

19 Kula, p. 12.

20 Luke 6:38.

21 Kula, p. 49.

22 James C. Scott, *The Moral Economy of the Peasant: Rebellion and Subsistence in Southeast Asia* (New Haven and London: Yale University Press, 1977), p. 71.

23 Kula, p. 189.

24 Gilbert Shapiro and John Markoff, *Revolutionary Demands: A Content Analysis of the Cahiers de Doléances of 1789* (Stanford, CA: Stanford University Press, 1998), p. 381.

25 Gerhard Dohrn-van Rossum, *History of the Hour: Clocks and Modern Temporal Order*, Thomas Dunlap (trans.) (Chicago and London: University of Chicago Press, 1996), p. 19.

26 Eighteen total decans would appear in the night, but only twelve were counted during total darkness, marking a period of forty minutes apiece. Sebastian Porceddu et al., 'Algol as Horus in the Cairo Calendar: The

Possible Means and the Motives of the Observations', *Open Astronomy*, 27, 2018, pp. 232–64.

27 Edoardo Detoma, 'On Two Star Tables on the Lids of Two Coffins in the Egyptian Museum of Turin', *Archeologia, Epigrafia e Numismatica*, 2014, pp. 117–69.

28 Jean Gimpel, *The Medieval Machine: The Industrial Revolution of the Middle Ages* (Harmondsworth: Penguin Books, 1976 (1983)), p. 168.

29 Quoted in Alfred W. Crosby, *The Measure of Reality: Quantification and Western Society*, 1250–1600 (Cambridge: Cambridge University Press, 1997 (1998)), p. 32.

30 'Seven times a day do I praise thee because of thy righteous judgments', Psalms 119:164.

31 David S. Landes, *Revolution in Time: Clocks and the Making of the Modern World* (Cambridge, MA, and London: Belknap Press of Harvard University Press, 1983), pp. 404–5.

32 Crosby, p. 33.

33 Kelly Wetherille, 'Japanese Watchmaker Adapts Traditional Timepiece', *The New York Times*, 12 November 2015. https://www.nytimes.com/2015/11/12/fashion/japanese-watchmaker-adapts-traditional-timepiece.html.

34 'The World of Japanese Traditional Clock', Japan Clock & Watch Association. https://www.jcwa.or.jp/en/etc/wadokei.html.

35 Llewellyn Howes, '"Who Will Put My Soul on the Scale?" Psychostasia in Second Temple Judaism', *Old Testament Essays*, 27(1), 2014, pp. 100–22. Retrieved 14 August 2020 from http://www.scielo.org.za/scielo.php?script=sci_arttext&pid=S1010-99192014000100007&lng=en&tlng=en.

36 Samuel G. F. Brandon, 'The Weighing of the Soul', in Joseph M. Kitagawa and Charles H. Long (eds), *Myths and Symbols: Studies in Honor of Mircea Eliade* (Chicago: Chicago University Press, 1969), pp. 98–9.

37 B. C. Dietrich, 'The Judgement of Zeus', *Rheinisches Museum für Philologie*, Neue Folge, 107. Bd., 2. H. (1964), pp. 97–125.

38 I'm indebted to Emanuele Lugli for surfacing this fact (Lugli, p. 111), which he found in Alain Guerreau's 'Mensura et Metiri dans la Vulgate'. I performed my own tallies and found that the King James Version mentions 'charity' as a noun twenty-eight times, each time recommending it to the reader.

39 Quoted in Lugli, p. 155.

40 Quoted in Lugli, p. 147.

41 Wayland D. Hand, 'Measuring and Plugging: The Magical Containment

and Transfer of Disease', *Bulletin of the History of Medicine*, 48(2), Summer 1974, pp. 221–33. Harry Gray, W. I. Feagans, and Ernest W. Baughman, 'Measuring for Short Growth', *Hoosier Folklore*, 7(1), March 1948, pp. 15–19. Ellen Powell Thompson, 'Folk-Lore from Ireland', *The Journal of American Folklore*, 7(26), July–September 1894, pp. 224–7. Tom Peete Cross, 'Witchcraft in North Carolina', *Studies in Philology*, 16 (3), July 1919, pp. 217–87.

42 Lugli, p. 146.

43 Harry A. Miskimin, 'Two Reforms of Charlemagne? Weights and Measures in the Middle Ages', *The Economic History Review*, New Series, 20(1), April 1967, pp. 35–52.

44 Alexis Jean Pierre Paucton, *Métrologie ou Traité des Mesures, Poids et Monnoies des anciens Peuples & des Modernes* (La Veuve Desaint, 1780); via Kula, p. 163.

45 Lugli, pp. 88–90.

46 Patrick Boucheron, '"Turn Your Eyes to Behold Her, You Who Are Governing, Who Is Portrayed Here", Ambrogio Lorenzetti's Fresco of Good Government', *Annales. Histoire, Sciences Sociales*, 60(6), 2005, pp. 1,137–99.

47 Lugli, p. 196.

48 Lugli, p. 203.

49 Diana Wood, *Medieval Economic Thought* (Cambridge: Cambridge University Press, 2002), p. 91.

50 Brian A. Sparkes, 'Measures, Weights, and Money', in Edward Bispham, Thomas Harrison, and Brian A. Sparkes (eds), *The Edinburgh Companion to Ancient Greece and Rome*, pp. 471–6.

51 Alison E. Cooley, *Pompeii: A Sourcebook* (Abingdon: Routledge, 2004), p. 179.

52 Mabel Lang and Margaret Crosby, 'Weights, Measures and Tokens', *The Athenian Agora*, 10, 1964 (The American School of Classical Studies at Athens), pp. 1–146.

53 Dennis Romano, *Markets and Marketplaces in Medieval Italy, c.1100 to c.1400* (New Haven, CT: Yale University Press, 2015), pp. 217–19. From Lugli, p. 67.

54 Lugli, p. 67.

55 Lugli, p. 70.

56 Edward Nicholson, *Men and Measures: A History of Weights and Measures, Ancient and Modern* (London: Smith, Elder & Co., 1912), p. 60.

57 Lugli, p. 95.

58 From William Blackstone's 'Commentaries on the Laws of England',

quoted in Charles Gross, 'The Court of Piepowder', *The Quarterly Journal of Economics*, 20(2), February 1906, pp. 231–49.

59 Herbert Arthur Klein, *The Science of Measurement: A Historical Survey* (New York: Dover Publications, 1974), pp. 65–7.

CHAPTER 3: THE PROPER SUBJECT OF MEASUREMENT

1 James Spedding, Robert Leslie Ellis, and Douglas Denon Heath (eds), *The Works of Francis Bacon*, vol. 4, book I, aphorism 6 (Boston: Houghton, Mifflin & Co., 1858).

2 Edward Mendelson (ed.), *The English Auden: Poems, Essays and Dramatic Writings, 1927–1939* (London: Faber & Faber, 1986), p. 292. Quoted in Crosby, p. 12.

3 Alexandre Koyré, *From the Closed World to the Infinite Universe* (Baltimore: Johns Hopkins Press, 1957), p. 1.

4 John Donne, 'An Anatomy of the World' (1611), in Roy Booth (ed.), *The Collected Poems of John Donne* (Wordsworth Poetry Library, 1994), p. 177.

5 Aristotle, *Physics*, Book II.3, 194 b 16 – 194 b 23; Jonathan Barnes (ed.), *The Complete Works of Aristotle: The Revised Oxford Translation* (Princeton: Princeton University Press, 1984 (1991)).

6 Saint Augustine, *Confessions*, XIII.9, F. J. Sheed (trans.) (1942–3), introduction by Peter Brown (Indianapolis: Hackett Publishing, 2006), p. 294.

7 Nicholas of Cusa, *De docta ignorantia* II.13, Jasper Hopkins (trans.) (Minneapolis: Arthur J. Banning Press, 1985 (1990)), copyright 1981, p. 99.

8 From *Idiota de sapientia*, as quoted and translated in Charles Trinkaus, *The Scope of Renaissance Humanism* (Michigan: University of Michigan Press, 1983); from the essay 'Humanism and Greek Sophism: Protagoras in the Renaissance', p. 176.

9 Trinkaus, p. 176.

10 Crosby, p. 12.

11 Anne Carson, *Autobiography of Red* (London: Jonathan Cape, 1999 (2010)), p. 4.

12 R. C. Cross and A. D. Woozley, *Republic*, vii.522C; from *Plato's Republic: A Philosophical Commentary* (New York: Macmillan, 1964), p. 155.

13 Gregory Vlastos, *Plato's Universe* (Oxford: Clarendon Press, 1975), p. 97.

14 Plato, *Plato's Examination of Pleasure: A Translation of the Philebus*, introduction and commentary by R. Hackforth (Cambridge: Cambridge University Press, 1945), 56B, p. 117.

15 *De libero arbitrio*, book 2, chapter 8, section 21; quoted in A. C. Crombie, *The History of Science from Augustine to Galileo* (New York: Dover

Publications, 1995) (unabridged, revised, enlarged edn of the work first published in 1952), p. 33.

16 Carl B. Boyer, *A History of Mathematics* (Princeton: Princeton University Press, 1985), p. 96. In the *Republic*, Plato also claims that a good king will live 729 (9 cubed) times more pleasantly than a tyrant. As Thomas Ainsworth, philosophy lecturer at Lady Margaret Hall and Trinity College Oxford, told me: 'My own view is that these are rare Platonic jokes.'

17 A. C. Crombie, *Styles of Scientific Thinking in the European Tradition*, vol. 1 (London: Gerald Duckworth & Co. Ltd, 1994), p. 99. Crombie: attaching higher symbolic importance to numbers 'made mathematics thus an instrument at once of calculation and of scientific research into the nature of things'.

18 Faith Wallis, '"Number Mystique" in Early Medieval Computus Texts', in T. Koetsier and L. Bergmans (eds), *Mathematics and the Divine: A Historical Study* (Amsterdam: Elsevier B.V., 2005), p. 182.

19 Vincent Foster Hopper, *Medieval Number Symbolism: Its Sources, Meaning, and Influence on Thought and Expression* (New York: Columbia University Press, 1938), pp. 94–5.

20 Kings 20:30, KJV.

21 Crosby, p. 27.

22 Edward Gibbons, *The History of the Decline and Fall of the Roman Empire* (New York: Harper & Brothers Publishers, 1879. Originally published 1776), p. 418.

23 Stephen A. Barney, W. J. Lewis, J. A. Beach, Oliver Berghof, with the collaboration of Muriel Hall, *The Etymologies of Isidore of Seville* (Cambridge: Cambridge University Press, 2006), XII.iii.10–iv.11, p. 255.

24 Barney et al., II.iv.1–v.8, p. 90.

25 All quotes from Edward Grant, *The Foundations of Modern Science in the Middle Ages* (Cambridge: Cambridge University Press, 1996 (1998)), pp. 68–9.

26 Aristotle, *Posterior Analytics*, I.2, 71b9–11; II.11, 94a20, in Jonathan Barnes (ed.), *Complete Works of Aristotle, Volume 1: The Revised Oxford Translation* (Princeton: Princeton University Press, 1984), book I, chapter 18, p. 132.

27 Steven Shapin, *The Scientific Revolution* (Chicago and London: University of Chicago Press, 1996 (1998)), p. 29.

28 Edith Dudley Sylla, 'The Oxford Calculators', in Norman Kretzmann, Anthony Kenny, and Jan Pinborg (eds), *The Cambridge History of Later Medieval Philosophy* (Cambridge: Cambridge University Press, 1982),

pp. 540–63. Significant works include Thomas Bradwardine's *De proportionibus velocitatum in motibus* (1328); William Heytesbury's *Regulae solvendi sophismata* (1335); and Richard Swineshead's *Liber calculationum* (c.1350).

29 Thomas Bradwardine, *Tractatus de continuo*, quoted by J. A. Weisheipl, 'Ockham and the Mertonians', in T. H. Aston (ed.), *The History of the University of Oxford* (Oxford: Clarendon Press, 1984), pp. 607–58, p. 627.

30 Mark Thakkar, 'The Oxford Calculators', in *Oxford Today: The University Magazine*, Trinity Issue 2007, pp. 24–6.

31 Weisheipl, pp. 607–58.

32 Edgar Zilsel, *The Social Origins of Modern Science*, Diederick Raven, Wolfgang Krohn, and Robert S. Cohen (eds), *Boston Studies in the Philosophy and History of Science*, vol. 200 (Dordrecht, Boston and London: Kluwer Academic Publishers, 2000), p. 4.

33 As in Bonaventura Berlinghieri's thirteenth-century altarpiece depicting the life of Francis of Assisi, which has panels showing action happening inside and outside a church at the same time.

34 Quoted in Frank Prager and Gustina Scaglia, *Mariano Taccola and His Book 'De Ingeneis'* (Cambridge, MA, and London: MIT Press, 1972), p. 11.

35 Alberti wrote several editions, one in Tuscan vernacular (printed in Italian as *Della Pittura*) and one in Latin (*De pictura*), before revising and re-editing these multi-volume texts over the next 30 years. While the vernacular version seems to have been written first, it's the Latin edition that spread most widely.

36 Leon Battista Alberti, *On Painting and On Sculpture: The Latin Texts of 'De Pittura' and 'De Statua'*, Cecil Grayson (trans.) (London: Phaidon, 1972) 55, pp. 67–9.

37 A. Mark Smith, 'The Alhacenian Account of Spatial Perception and Its Epistemological Implications', *Arabic Sciences and Philosophy*, 15(2) (Cambridge: Cambridge University Press, 2005), p. 223.

38 Quoted in Samuel Y. Edgerton, *The Mirror, the Window, and the Telescope* (New York: Cornell University Press, 2009), p. 89.

39 *The Etymologies*, III.xiv.4–xvii.3 p. 95.

40 Hugo Riemann, *History of Music Theory: Books I and II, Polyphonic Theory to the Sixteenth Century* (Lincoln, NE: University of Nebraska Press, 1962), pp. 131–57 ('Mensural Theory to the Beginning of the 14th Century').

41 See Max Weber, in Don Martindale, Johannes Riedel, and Gertrude Neuwirth (eds), *The Rational and Social Foundations of Music* (Carbondale: Southern Illinois University Press, 1958), pp. 82–8.

42 Strunk's *Source Readings in Music History*, 1 (New York: Norton, 1989), pp. 184–5, 189, 190; Craig Wright, *Music and Ceremony at Notre Dame of Paris, 500–1550* (Cambridge: Cambridge University Press, 1989), p. 345. Quoted in Crosby, p. 158.

43 Quoted in Henry Raynor, *A Social History of Music from the Middle Ages to Beethoven* (London: Barrie & Jenkins, 1972), pp. 36–7.

44 Landes, pp. 6–11.

45 For example, the astronomical dial built by Roger Stoke for Norwich Cathedral (1321–5); and Richard of Wallingford's astronomical mechanism, installed at St Albans, which was completed in 1364 after 30 years of work. Giovanni de Dondi's astronomical clock, completed in around 1380, was the secular marvel of its time.

46 For proponents of this theory, see Lewis Mumford, who says Benedictines are the founders of modern capitalism as they helped to give 'human enterprise the regular collective beat and rhythm of the machine', and H. E. Hallam, who says the 'spirit of the clock' is 'wholly Benedictine'. For a critique of this interpretation, see Gerhard Dohrn-van Rossum's *History of the Hour*.

47 Quoted in Landes, p. 65.

48 Gerhard Dohrn-van Rossum, p. 38. Note: Rossum is paraphrasing the arguments of Arno Borst.

49 Lynn White, *Medieval Technology and Social Change* (Oxford: Clarendon Press, 1962), p. 124.

50 Quoted in Carlo M. Cipolla, *Clocks and Culture 1300–1700* (London: Collins, 1967), p. 42.

51 Landes, p. 81.

52 Lewis Mumford, *Technics and Civilization* (London: Routledge, 1923 (1955)), pp. 13–14.

53 Quoted in Shapin, p. 32.

54 From Galileo's *Sidereus Nuncius*; quoted in Koyré, p. 89.

55 Quoted in Shapin, p. 18.

56 Quoted in Shapin, p. 33.

57 Shapin, p. 62.

58 Max Weber, *The Vocation Lectures*, David Owen and Tracy B. Strong (eds), Rodney Livingstone (trans.) (Indianapolis and Cambridge: Hackett Publishing Company, 2004).

59 Quoted in Shapin, p. 63.

60 John Maynard Keynes, *Essays in Biography, Vol. 10: The Collected Writings of John Maynard Keynes* (Palgrave Macmillan/Royal Economic Society, 1972), p. 363.

CHAPTER 4: THE QUANTIFYING SPIRIT

1 Robert Frost, '"Fire and Ice", A Group of Poems by Robert Frost', *Harper's Magazine*, vol. 142, December 1920, p. 67.

2 Tore Frängsmyr, J. L. Heilbron, and Robin E. Rider (eds), *The Quantifying Spirit in the 18th Century* (Berkeley: University of California Press, 1990).

3 Francis Bacon, *New Atlantis* (1627), from Susan Ratcliffe (ed.), *Oxford Essential Quotations* (Oxford: Oxford University Press, 2016 (online version), fourth edn).

4 Robert I. Frost, *The Northern Wars: War, State and Society in Northeastern Europe 1558–1721* (London: Longman, 2000), pp. 133–4.

5 Robert I. Frost, *After the Deluge: Poland–Lithuania and the Second Northern War, 1655–1660* (Cambridge: Cambridge University Press, 1993).

6 Martin Ekman, *The Man Behind 'Degrees Celsius': A Pioneer in Investigating the Earth and Its Changes* (Summer Institute for Historical Geophysics, Åland Islands, 2016).

7 See Hasok Chang, *Inventing Temperature: Measurement and Scientific Progress* (Oxford: Oxford University Press, 2004). And interviews with the author.

8 Quoted in Constantine J. Vamvacas, *The Founders of Western Thought – The Presocratics*, Robert Crist (trans.), *Boston Studies in the Philosophy and History of Science*, vol. 257 (Dordrecht: Springer, 2009), p. 119.

9 Richard J. Durling, 'The Innate Heat in Galen', *Medizinhistorisches Journal*, bd. 23, h. 3/4, 1988, pp. 210–12.

10 W. E. Knowles Middleton, *A History of the Thermometer and Its Use in Meteorology* (Baltimore: Johns Hopkins University Press, 1966), pp. 3–5.

11 Quoted in W. E. Knowles Middleton, *Catalog of Meteorological Instruments in the Museum of History and Technology* (Washington, DC: Smithsonian Institution Press, 1969), p. 37.

12 Fabrizio Bigotti and David Taylor, 'The Pulsilogium of Santorio: New Light on Technology and Measurement in Early Modern Medicine', *Soc. Politica*, 11(2), 2017, pp. 53–113.

13 Quoted in Teresa Hollerbach, 'The Weighing Chair of Sanctorius Sanctorius: A Replica', *NTM*, 26(2) 2018, pp. 121–49.

14 Quoted in Middleton, 1966, p. 7.

15 Martin K. Barnett, 'The Development of Thermometry and the Temperature Concept', *Osiris*, 12, 1956, p. 277.

16 Francis Bacon, *Advancement of Learning and Novum Organum* (London and New York: Colonial Press, 1900), p. 387.

17 Middleton, 1966, p. 20.

18 From Jean Leurechon's *Récréations mathématiques* (1624), quoted in Henry Carrington Bolton, *Evolution of the Thermometer 1592–1743* (Easton, PA: The Chemical Publishing Co., 1900), pp. 11–12.

19 Original: Cornelis van der Woude and Pieter Jansz Schaghen, *Kronyck van Alckmaar* (Amsterdam: Steven van Esveldt, 1742), p. 102. Translation: Hubert van Onna, Drebbologist and Chairman of the Second Drebbel Foundation, 'Cornelis Jacobszoon Drebbel, "a Bold mind, a show-off to the World"'. www.drebbel.net.

20 Gerrit Tierie, *Cornelis Drebbel (1572–1633)* (Amsterdam: H. J Paris, 1932). http://www.drebbel.net/Tierie.pdf.

21 Dr James Bradburne, 'Afbeeldingen van Cornelis Drebbel's Perpetuum Mobile', January 2015. http://www.drebbel.net/Drebbels%20 Perpetuum%20Mobile.pdf

22 Michael John Gorman, *Mysterious Masterpiece: The World of the Linder Gallery* (Firenze, Italy: Mandragora Srl/Alias, 2009).

23 Edmond Halley, 'An Account of Several Experiments Made to Examine the Nature of the Expansion and Contraction of Fluids by Heat and Cold, in Order to Ascertain the Divisions of the Thermometer, and to Make That Instrument, in All Places, Without Adjusting by a Standard', *Philosophical Transactions*, 17(197), 1693 (The Royal Society), p. 655.

24 Suggested by the Accademia del Cimento in the mid-1600s and by Joachim Dalencé in 1688 respectively. See Chang, p. 10.

25 I. Bernard Cohen (ed.), *Isaac Newton's Papers and Letters on Natural Philosophy* (Cambridge, MA: Harvard University Press, 1958), pp. 265–8.

26 Pieter van der Star (ed.), *Daniel Gabriel Fahrenheit's Letters to Leibniz and Boerhaave* (Museum Boerhaave, 1983), p. 3.

27 From *Elementa Chemiae*, a textbook of chemistry first published by Herman Boerhaave in 1732; cited in Chang, p. 58.

28 Chang, p. 12.

29 Chang, p. 11.

30 Chang, p. 45.

31 William Thomson, 'On an Absolute Thermometric Scale', *Philosophical Magazine*, 1848, from Sir William Thomson, *Mathematical and Physical Papers*, vol. 1 (Cambridge: Cambridge University Press, 1882); quoted in Chang, p. 159.

32 Quoted in Gaston Bachelard, *The Psychoanalysis of Fire* (London: Routledge & Kegan Paul, 1964), p. 60.

33 Antoine Laurent Lavoisier, *Elements of Chemistry*, Robert Kerr (trans.) (New York: Dover Publications, 1965. Original: Edinburgh: William Creech, 1790), pp. 3–7.

34 William Thomson, 'An Account of Carnot's Theory of the Motive Power of Heat; With Numerical Results Deduced from Regnault's Experiments on Steam', in *Mathematical and Physical Papers*, Cambridge Library Collection – Physical Sciences (Cambridge: Cambridge University Press, 2011), pp. 100–6.

35 David Lindley, *Degrees Kelvin: A Tale of Genius, Invention, and Tragedy* (Washington, DC: Joseph Henry Press, 2004), p. 69.

36 P. W. Atkins, *The Second Law* (New York: Scientific American Books, 1984), p. 1.

37 Lindley, p. 75.

38 Quoted in Lindley, p. 75.

39 Quoted in Wayne M. Saslow, 'A History of Thermodynamics: The Missing Manual', *Entropy*, 22(1), 77, 2020, p. 21.

40 Rudolf Clausius, *The Mechanical Theory of Heat, with Its Applications to the Steam-Engine and to the Physical Properties of Bodies*, Thomas Archer Hirst (ed.), John Tyndall (trans.) (London: John van Voorst, 1867), p. 386.

41 Porter, *Trust in Numbers*, p. 18.

42 Quoted in Frängsmyr et al., 1990, p. 1.

43 John Smith, 'The Best Rules for the Ordering and Use Both of the Quick-Silver and Spirit Weather-Glasses', fl. 1673–80. https://quod.lib.umich.edu/e/eebo/A60473.0001.001.

44 'The Female Thermometer', in Terry Castle, *The Female Thermometer: Eighteenth-Century Culture and the Invention of the Uncanny* (Oxford: Oxford University Press, 1995), p. 27.

45 Clausius, p. 357.

46 William Thomson, 'On the Age of the Sun's Heat', *Macmillan's Magazine*, vol. 5 (5 March 1862), pp. 388–93.

47 See Crosbie Smith, 'Natural Philosophy and Thermodynamics: William Thomson and "The Dynamical Theory of Heat"', *The British Journal for the History of Science*, 9(3), 1976, pp. 293–319.

48 Crosbie Smith and M. Norton Wise, *Energy and Empire: A Biographical Study of Lord Kelvin* (New York: Cambridge University Press, 1989), pp. 331, 535.

49 Barri J. Gold, 'The Consolation of Physics: Tennyson's Thermodynamic Solution', *PLMA*, 117(3), May 2002, pp. 449–64 (p. 452).

50 This argument I credit to Dr Elizabeth Neswald of Brock University and her enlightening discussions with me.

51 Tamara Ketabgian, 'The Energy of Belief: The Unseen Universe and the Spirit of Thermodynamics', in *Strange Science: Investigating the Limits of*

Knowledge in the Victorian Age (Ann Arbor, MI: University of Michigan Press, 2017), p. 255.

52 Algernon Charles Swinburne, 'The Garden of Proserpine', in *Poems and Ballads* (London: James Camden Hotten, 1886), pp. 196–9.

53 Quoted in Suzy Anger, 'Evolution and Entropy: Scientific Contexts in the Nineteenth Century', in Robert DeMaria Jr., Heesok Chang, and Samantha Zacher (eds), *A Companion to British Literature: Part IV: Victorian and Twentieth-Century Literature 1837–2000* (Hoboken, NJ: John Wiley & Sons, 2014), p. 62.

54 H. G. Wells, *The Time Machine* (1895).

CHAPTER 5: THE METRIC REVOLUTION

1 'Outlines of an Historical View of the Progress of the Human Mind, being a posthumous work of the late M. de Condorcet'. Translated from the French (Philadelphia: M. Carey, 1796), p. 259.

2 Michael Trott (trans.), 'As of Today, the Fundamental Constants of Physics (c, h, e, k, N_A) Are Finally . . . Constant!' Wolfram blog: https://blog.wolfram.com/2018/11/16/as-of-today-the-fundamental-constants-of-physics-c-h-e-k-na-are-finally-constant/. Original from Laplace's speech to the Council of Five Hundred, sourced in Jean-Baptiste Delambre and Pierre Méchain, 'Base du système métrique décimal, ou Mesure de l'arc du méridien compris entre les parallèles de Dunkerque et Barcelone', 1806–10, Bibliothèque nationale de France, département Réserve des livres rares, V-7586. https://gallica.bnf.fr/ark:/12148/bpt6k1106055.

3 Quoted in Lugli, p. 37.

4 K. M. Delambre, *Base du système du métrique décimal*, 1: title page (Paris: Baudouin, 1806, 1807, 1810). As quoted in Ken Alder's *The Measure of All Things: The Seven-Year Odyssey and Hidden Error That Transformed the World* (New York: Free Press, 2002 (2003)), p. 3.

5 Alder, p. 7.

6 Arthur Young, *Travels During the Years 1787, 1788, and 1789: Undertaken More Particularly with a View of Ascertaining the Cultivation, Wealth, Resources, and National Prosperity of the Kingdom of France*, vol. 1 (London: printed by J. Rackham for W. Richardson, 1792), p. 302.

7 J. L. Heilbron, 'The Measure of Enlightenment', in Frängsmyr et al., p. 207.

8 Roland Edward Zupko, *Revolution in Measurement: Western European Weights and Measures Since the Age of Science* (Memoirs of the American Philosophical Society), vol. 186 (American Philosophical Society, 1990), p. 113.

9 John Markoff, 'Peasants Protest: The Claims of Lord, Church, and State in the *Cahiers de Doléances* of 1789', *Comparative Studies in Society and History*, 32(3), July 1990, pp. 413–54.

10 Kula, p. 192.

11 Kula, pp. 203–5. See, e.g., the *cahier* for Pas-de-Calais, which includes the phrase: 'And now we call upon the King to mete out justice, and we express a most sincere desire for but one king, one law, one weight, and one measure.'

12 From Sir John Riggs Miller, 'A Proposition Offered to the National Assembly on Weights and Measures by the Bishop of Autun', *Speeches in the House of commons upon the equalization of the weights and measures of Great Britain; with notes, &c. Together with two letters from the bishop of Autun* (London: J. Debrett, 1790), p. 77.

13 Talleyrand, *Archives parlementaires de 1787 à 1860; recueil complet des débats législatifs et politiques des chambres françaises* (Paris: Dupont, 9 March 1790), p. 106. Quoted in Alder, p. 85, who notes that 'identical phrasing can be found in Condorcet's proposal'.

14 John Riggs Miller, 'Speeches in the House of Commons upon the equalization of the weights and measures of Great Britain . . . Together with two letters from the Bishop of Autun to the author upon the uniformity of weights and measures; that prelate's proposition respecting the same to the National Assembly; and the decree of that body . . . With English translations', London, 1790; General Reference Collection of the British Library E.2159.(3).

15 Louis Jourdan, *La Grande Métrication* (Nice: France Europe Éditions, 2002). Original quote in French: '*Parfois en trahissant les rois ou les empereurs, mais jamais la France.*'

16 David Lawday, *Napoleon's Master: A Life of Prince Talleyrand* (London: Jonathan Cape, 2006), p. 2.

17 Auguste-Savinien Leblond, *Sur la fixation d'une mesure et d'un poid – lu à l'Académie des Sciences le 12 mai 1790* (Paris: Demonville, 1791), 10; quoted in Alder, p. 87.

18 Mohammed Abattouy, 'The Mathematics of Isochronism in Galileo: From His Manuscript Notes on Motion to the Discorsi', *Societate si Politica*, 11(2), January 2017, pp. 23–54.

19 J. Donald Fernie, 'Marginalia: The Shape of the Earth', *American Scientist*, 79(2), March–April 1991, pp. 108–10. For Lignes measurement, see John Henry Poynting and Joseph John Thompson, *A Textbook of Physics: Properties of Matter*, fourth edn (London: Charles Griffin & Co., 1907), p. 20.

20 John L. Greenberg, 'Isaac Newton and the Problem of the Earth's Shape', *Archive for History of Exact Sciences*, 49(4), 1996, pp. 371–91.

21 Alder, p. 3.

22 Alder, p. 31.

23 Alder, p. 34.

24 Alder, p. 251.

25 Letter from Méchain to Rolland, 22 floréal VII (11 May 1799), in Dougados, 'Lettres de l'astronome Méchain à M. Rolland', *Mémoires de la Société des arts et des sciences de Carcassonne 2* (1856), 101, quoted in Alder, p. 250.

26 Alder, p. 252.

27 Alder, p. 303.

28 Ken Alder, 'A Revolution to Measure: The Political Economy of the Metric System in France', in M. Norton Wise (ed.), *The Values of Precision* (Princeton: Princeton University Press, 1995), pp. 39–71, p. 52.

29 Quoted in Alder, p. 90.

30 Alexander Pope, 'Epitaph: Intended for Sir Isaac Newton' (1730), from Ratcliffe *Oxford Essential Quotations*.

31 As quoted in Alexandre Koyré, 'Condorcet', *Journal of the History of Ideas* 9(2), April 1948, p. 135; p. 139 for the 'indefinite perfectibility' (all from *Esquisse d'un tableau historique des progrès de l'esprit humain* – 'Sketch for a Historical Picture of the Progress of the Human Mind'). Koyré's essay gives a fine overview of Condorcet's thought and its historical context.

32 Quoted in Steven Lukes and Nadia Urbinati (eds), *Condorcet: Political Writings* (Cambridge: Cambridge University Press, 2012), p. xviii.

33 Condorcet, 'Sketch for a Historical Picture of the Progress of the Human Mind: Tenth Epoch', Keith Michael Baker (trans.), *Dædalus*, 133(3), Summer 2004, p. 69.

34 Condorcet, 'Sketch', p. 78.

35 Condorcet, 'Sketch', p. 72.

36 Quoted in Keith Michael Baker, *Condorcet: From Natural Philosophy to Social Mathematics* (Chicago and London: University of Chicago Press, 1975), p. 367. Baker says it comes from a 'fragment related to the Tenth Epoch' (p. 366).

37 These classifications are purely a Linnaean hotchpotch of the author's own devising. The high-minded Condorcet's example was of a demographic table that would yield a 'veritable natural history of man'. (Quote also from Baker, 1975, p. 123.)

38 'Moreover, he clearly regarded it as a step towards the fulfillment of his own methodological program: that of creating a universal language of the

sciences that would provide the moral and political sciences with a logic as precise and analytical as that enjoyed by the physical sciences', Baker, p. 277.

39 Condorcet, *Note 1er Epoque X. Exemple des méthodes techniques,* quoted (and translated) in Baker, 1975, p. 123.

40 Antoine Laurent Lavoisier, *Elements of Chemistry in a New Systematic Order* (Edinburgh: William Creech, 1790), pp. 295–6. Quoted in Frängsmyr et al., p. 212.

41 Quoted by Alder, 'A Revolution to Measure', in *The Values of Precision,* p. 41.

42 Karl Marx, *Capital,* vol. 1 (1867), Ben Fowkes (trans.) (London: Penguin, 1976 (1982)), p. 644.

43 Maurice Crosland, '"Nature" and Measurement in Eighteenth-Century France', in *Studies on Voltaire and the Eighteenth Century, LXXXVII* (Geneva: The Voltaire Foundation, 1972), p. 285.

44 Alder, 2002, p. 135.

45 *Cahiers de doléances du bailliage d'Orléans pour les États généraux de 1789,* 2 vols, vol. 1, *Collection de documents inédits sur l'histoire économique de la Révolution française* (Orléans, 1906), p. 615, 'sont les sangsues de la Nation et boivent dans des coupes d'or les pleurs des malheureux'.

46 Quoted in William F. McComas (ed.), *Nature of Science in Science Instruction: Rationales and Strategies* (Cham: Springer Nature, 2020), p. 567.

47 Alder, 2002, p. 137.

48 From John Riggs Miller, an English MP who corresponded with Talleyrand on the metric project and campaigned unsuccessfully for English involvement. Quoted in Sally Riordan, 'Le Grave: The First Determination of the Kilogram, 1789–1799', a dissertation submitted to the department of philosophy and the committee on graduate studies of Stanford University in partial fulfilment of the requirements for the degree of Doctor of Philosophy (May 2012), pp. 129–30. As sent to the author.

49 Maximilien Robespierre, in Lewis Copeland, Lawrence W. Lamm, and Stephen J. McKenna (eds), *The World's Great Speeches,* fourth edn (New York: Dover Publications, 1999), p. 84.

50 Jonathan Smyth, *Robespierre and the Festival of the Supreme Being* (Manchester: Manchester University Press, 2016), p. 58.

51 The quote: 'Ever since the sun has stood in the heavens, and the planets revolved around it, never have we known man to walk on his head, that is, to base himself on the Idea and to build the world in accordance with it.'

52 David Andress, *The Terror: Civil War in the French Revolution* (London: Abacus, New Edition, 2005 (2006)), p. 291.

53 Matthew Shaw, *Time and the French Revolution: The Republican Calendar, 1789–Year XIV* (Rochester, NY: Boydell Press/The Royal Historical Society, 2011), p. 61.

54 P. F. Fabre d'Églantine, *Rapport fait à la Convention nationale au nom de la Commission chargée de la confection du Calendrier* (Paris, Imprimerie nationale, 1793). Quoted in Brendan Dooley, 'The Experience of Time', in Alessandro Arcangeli, Jörg Rogge, and Hannu Salmi (eds), *The Routledge Companion to Cultural History in the Western World* (London and New York: Routledge, 2020).

55 John Brady, *Clavis Calendaria: Or, A Compendious Analysis of the Calendar; Illustrated with Ecclesiastical, Historical, and Classical Anecdotes*, vol. 1 (London: Rogerson and Tuxford, 1812), p. 39.

56 Shaw, pp. 17–28.

57 Mona Ozouf, *Festivals and the French Revolution*, Alan Sheridan (trans.) (Cambridge, MA: Harvard University Press, 1988), p. 2.

58 Ozouf, p. 13.

59 Shaw, p. 83.

60 Shaw, p. 145. See also Sanja Perovic, *The Calendar in Revolutionary France: Perceptions of Time in Literature, Culture, and Politics* (Cambridge: Cambridge University Press, 2012).

61 Shaw, p. 52.

62 Alder, p. 260.

63 Alder, p. 257.

64 Quoted in Heilbron, 'The Measure of Enlightenment', p. 238.

65 Napoleon, *Mémoires*, 4:211–15. Quoted in Alder, p. 318.

CHAPTER 6: A GRID LAID ACROSS THE WORLD

1 Edward W. Said, *Culture and Imperialism* (New York: Vintage Books, 1993 (1994)), p. 78.

2 James C. Scott, *Seeing Like a State: How Certain Schemes to Improve the Human Condition Have Failed* (New Haven and London: Yale University Press, 1999), p. 3.

3 Scott, pp. 64–76.

4 William Camden, *Remains Concerning Britain*, R. D. Dunn (ed.) (1605; Toronto: University of Toronto Press, 1984, p. 122); cited in Scott, p. 372.

5 Quoted in Guy H. Dodge, *Benjamin Constant's Philosophy of Liberalism: A Study in Politics and Religion*, via Google Books (it's a different translation to Alder).

6 Quoted in Alder, p. 317.

7 King James Version.

8 Ovid, *Metamorphoses*, Loeb Classical Library, pp. 11–12.

9 Andrew McRae, 'To Know One's Own: Estate Surveying and the Representation of the Land in Early Modern England', *The Huntington Library Quarterly*, 56(4), Autumn 1993, pp. 333–57.

10 E. G. R. Taylor, 'The Surveyor', *The Economic History Review*, 17(2), 1947, pp. 121–33.

11 Quoted in Henry S. Turner, 'Plotting Early Modernity', in Henry Turner (ed.), *The Culture of Capital: Property, Cities, and Knowledge in Early Modern England* (London: Routledge, 2002), pp. 101–2. Note: three edns of *The Surveiors Dialogve* were published, in 1607, 1608, and 1618.

12 Quoted in Michael Houseman, 'Painful Places: Ritual Encounters with One's Homelands', *The Journal of the Royal Anthropological Institute*, 4(3) (September 1998), p. 450. Original source: J. S. Udal, *Dorsetshire Folklore* (St Peter Port, Guernsey, 1922; Toucan Press, 1970).

13 Source: https://www.parliament.uk/about/living-heritage/ evolutionofparliament/originsofparliament/birthofparliament/ overview/magnacarta/magnacartaclauses.

14 Sir Thomas Smith, *De Republica Anglorum* (written 1565, pub. 1583); quoted in Robert Bucholz and Newton Key, *Early Modern England 1485–1714: A Narrative History* (New Jersey: Wiley-Blackwell, 2003), p. 7.

15 Jeffrey Ostler, *Surviving Genocide: Native Nations and the United States from the American Revolution to Bleeding Kansas* (New Haven and London: Yale University Press, 2019), p. 85.

16 Thomas Jefferson, *Notes on the State of Virginia 1743–1826* (Philadelphia: Prichard and Hall, 1788), p. 175.

17 Jefferson, p. 175.

18 'From George Washington to Lafayette, 25 July 1785', Founders Online, National Archives. https://founders.archives.gov/documents/ Washington/04-03-02-0143. (Original source: *The Papers of George Washington*, Confederation Series, vol. 3, 19 May 1785–31 March 1786, W. W. Abbot (ed.) (Charlottesville: University Press of Virginia, 1994), pp. 151–5.)

19 Letter to Dr Percival, 4 April 1769. Benjamin Franklin, *The Works of Benjamin Franklin, Volume IV* (Philadelphia: William Duane, 1809), p. 206.

20 Richard White, *The Middle Ground: Indians, Empires, and Republics in the Great Lakes Region, 1650–1815* (New York: Cambridge University Press, 1991 (2011)), p. XXVI.

21 Alexis de Tocqueville, *Democracy in America*, Henry Reeve (trans.) (New York: Barnes & Noble, 2003), p. 268.

22 Paul Frymer, *Building an American Empire: The Era of Territorial and Political Expansion* (Princeton and Oxford: Princeton University Press, 2017), pp. 8–9.

23 The numbering system is idiosyncratic, with a coincidental connection to the land. Numbering starts in the top right, then sweeps horizontally left and right, back and forth, a pattern known as 'boustrophedonic', meaning 'ox-turning' in the way it follows the track of a plough.

24 Andro Linklater, *Measuring America: How an Untamed Wilderness Shaped the United States and Fulfilled the Promise of Democracy* (New York: HarperCollins, 2002), p. 75.

25 Quoted in Frederick J. Turner, 'The Problem of the West', *The Atlantic*, September 1896. https://www.theatlantic.com/magazine/archive/1896/09/the-problem-of-the-west/525699/.

26 C. Albert White, *A History of the Rectangular Survey System* (US Department of the Interior Bureau of Land Management, 1983), p. 29.

27 Linklater, pp. 184–5.

28 Ray Allen Billington and Martin Ridge, *Westward Expansion, A History of the American Frontier*, sixth edn (Albuquerque: University of New Mexico Press, 2001), p. 29.

29 Henry Clay, 'On Distributing the Proceeds of the Sales of the Public Lands Among the Several States' (16 April 1832), 22nd Congress, 1st Session, No. 1053.

30 Manasseh Cutler, *An Explanation of the Map Which Delineates That Part of the Federal Lands, Comprehended between Pennsylvania West Line, the Rivers Ohio and Sioto, and Lake Erie; Confirmed to the United States by Sundry Tribes of Indians, in the Treaties of 1784 and 1786, and Now Ready for Settlement* (Salem, MA: Dabney and Cushing, 1787), p. 21; quoted in Frymer, p. 55.

31 Charles Piazzi Smyth, *Our Inheritance in the Great Pyramid* (London: Alexander Strahan and Co., 1864), p. 372.

32 US Department of the Interior, Bureau of Land Management, *Public Land Statistics 2019*, vol. 204, June 2020, p. 1.

33 Letter to Major John Cartwright, Monticello, 5 June 1824. In *Memoirs, Correspondence and Private Papers of Thomas Jefferson, Late President of the United States, Volume IV* (London: Henry Colburn and Richard Bentley, 1829), p. 405.

34 Frances Trollope, *Domestic Manners of the Americans* (London: Richard Bentley, 1832 (1839)), p. 94.

35 Trollope, p. 39.

36 Harriet Martineau, *Society in America: Vol. 1* (Paris: Baudry's European Library, 1837), p. 203.

37 Quoted in Stuart Banner, *How the Indians Lost Their Land: Law and Power on the Frontier* (Cambridge, MA, and London: Belknap Press of Harvard University Press, 2005), p. 21.

38 Quoted in Ostler, pp. 201–2.

39 E.g. the Treaty of Fort Laramie in 1851; Billington and Ridge, p. 301.

40 John Heckewelder, *History, Manners, and Customs of the Indian Nations Who Once Inhabited Pennsylvania and the Neighbouring States* (Philadelphia: Abraham Small, 1819 (1881)), p. 336.

41 Walter Johnson, *River of Dark Dreams: Slavery and Empire in the Cotton Kingdom* (Cambridge, MA, and London: Belknap Press of Harvard University Press, 2013), p. 36.

42 Johnson, p. 41.

43 Keith H. Basso, *Wisdom Sits in Places* (Albuquerque: University of New Mexico Press, 1996 (1999)), p. 34.

44 Vine Deloria Jr., *God Is Red: A Native View of Religion* (Golden, CO: Fulcrum Publishing, 1973 (2003)), p. 61.

45 Alexis de Tocqueville, *Democracy in America*, J. P. Mayer (ed.) (New York: Perennial, 2000), p. 324.

46 In the US, for example, former slaves were promised that they would receive '40 acres and a mule' after the Civil War – an area defined by the smallest plot in the grid survey. This promise was broken, though, when white landowners demanded to keep their land (as well as a labour force with no other option but to work it).

47 Micheál Ó Siochrú and David Brown, 'The Down Survey and the Cromwellian Land Settlement', in Jane Ohlmeyer (ed.), *The Cambridge History of Ireland, Volume II, 1550–1730* (Cambridge: Cambridge University Press, 2018), pp. 584–607.

48 Adam Fox, 'Sir William Petty, Ireland, and the Making of a Political Economist, 1653–87', *The Economic History Review*, New Series, 62(2), May 2009, pp. 388–404.

49 J. G. Simms, 'The Restoration, 1660–85', in T. W. Moody, F. X. Martin, and F. J. Byrne (eds), *A New History of Ireland, III: Early Modern Ireland 1534–1691* (Oxford: Oxford University Press, 2009), p. 428.

50 W. J. Smyth, *Map-Making, Landscapes and Memory: A Geography of Colonial and Early Modern Ireland, c.1530–1750* (Cork: Cork University Press, 2006), p. 196.

51 Matthew H. Edney, *Mapping an Empire: The Geographical Construction*

of British India, 1765–1843 (Chicago: University of Chicago Press, 1990 (1997)), p. 2.

52 Hannah Arendt, *The Human Condition* (Chicago: University of Chicago Press, 1958 (1998)), pp. 250–1.

53 'Edgar Mitchell's Strange Voyage', *People*, 8 April 1974, #6, p. 20. https://people.com/archive/edgar-mitchells-strange-voyage-vol-1-no-6/.

54 Scott, p. 83.

55 Scott, p. 83.

CHAPTER 7: MEASURING LIFE AND DEATH

1 Louis-Sébastien Mercier, *Le nouveau Paris*, Volume 1 (London: H. D. Symonds, 1800), p. 324.

2 Thomas Sprat, *The History of the Royal-Society of London, for the Improving of Human Knowledge* (London: printed by T. R. for J. Martyn and J. Allestry, 1667); quoted in Ian Sutherland, 'John Graunt: A Tercentenary Tribute', *Journal of the Royal Statistical Society*, Series A (General), 126(4), 1963, p. 539.

3 Quoted in Sutherland, p. 552.

4 Quoted in Sutherland, p. 542.

5 K. J. Rothman, 'Lessons from John Graunt', *The Lancet*, 347 (8993), 6 January 1996, pp. 37–9. doi: 10.1016/s0140-6736(96)91562-7. PMID: 8531550.

6 Quoted in Sutherland, p. 541.

7 Quoted in Sutherland, p. 542.

8 Charles Henry Hull (ed.), *The Economic Writings of Sir William Petty, Together with the Observations upon the Bills of Mortality More Probably by Captain John Graunt*, vol. II (Cambridge: At the University Press, 1899), p. 554.

9 Hull, vol. I, p. 129.

10 Sutherland, p. 542.

11 Lorraine Daston, 'Why Statistics Tend Not Only to Describe the World but to Change It', *London Review of Books*, 22(8), April 2000.

12 Dirk Philipsen, *The Little Big Number: How GDP Came to Rule the World and What to Do about It* (Princeton: Princeton University Press, 2015).

13 Remarks at the University of Kansas, 18 March 1968. https://www.jfklibrary.org/learn/about-jfk/the-kennedy-family/robert-f-kennedy/robert-f-kennedy-speeches/remarks-at-the-university-of-kansas-march-18-1968.

14 Alain Desrosières, *The Politics of Large Numbers: A History of Statistical Reasoning* (Cambridge, MA: Harvard University Press, 1998), p. 9.

15 Paul F. Lazarsfeld, 'Notes on the History of Quantification in Sociology – Trends, Sources and Problems', *Isis*, 52(2), 1961, pp. 277–333.

16 Saul Stahl, 'The Evolution of the Normal Distribution', *Mathematics Magazine*, 79(2), 2006, pp. 96–113.

17 Quoted in Jed Z. Buchwald, 'Discrepant Measurements and Experimental Knowledge in the Early Modern Era', *Archive for History of Exact Sciences*, 60(6), November 2006, p. 32.

18 Stephen M. Stigler, *The History of Statistics: The Measurement of Uncertainty Before 1900* (Cambridge, MA, and London: Belknap Press of Harvard University Press, 1986), p. 21.

19 Quoted in Stigler, *The History of Statistics*, p. 27.

20 Stigler, p. 11.

21 Quoted in Theodore M. Porter, *The Rise of Statistical Thinking: 1820–1900* (Princeton: Princeton University Press, 1986), p. 44.

22 Ian Hacking, *The Taming of Chance* (Cambridge: Cambridge University Press, 1990), p. 2.

23 Hacking, p. 2.

24 Quoted in Porter, *The Rise of Statistical Thinking*, p. 103.

25 Desrosières, pp. 77–81.

26 *Athenaeum*, 29 August 1835, p. 661. Quoted in Stigler, p. 170.

27 Gerd Gigerenzer, Zeno Swijtink, Theodore Porter, Lorraine Daston, John Beatty, and Lorenz Krüger, *The Empire of Chance* (Cambridge: Cambridge University Press, 1989 (1997)), p. 129.

28 Quoted in Hacking, p. 105.

29 A. Quetelet, 'Sur la possibilité de mesurer l'influence des causes qui modifient les elements sociaux, Lettre à M. Villermé', *Correspondances mathématiques et physiques*, 7 (1832), p. 346. Quoted in Hacking, p. 114.

30 Hacking, p. 126.

31 Marion Diamond and Mervyn Stone, 'Nightingale on Quetelet', *Journal of the Royal Statistical Society*, Series A (General), 144(1), 1981, pp. 66–79.

32 Julian Wells, 'Marx Reads Quetelet: A Preliminary Report', September 2017. Online at https://mpra.ub.uni-muenchen.de/98255/. MPRA Paper No. 98255, posted 27 January 2020.

33 Henry Thomas Buckle, *The History of Civilization in England*, vol. I (Toronto: Rose-Belford Publishing Company, 1878), p. 121.

34 Buckle, p. 272.

35 Quotes from Hacking, pp. 13–14.

36 As quoted in Desrosières, p. 35.

37 Porter, *The Rise of Statistical Thinking*, p. 159.

38 Charles Dickens, *Hard Times and Reprinted Pieces* (London: Chapman & Hall, 1854 (1905)). Project Gutenberg eBook.

39 Laura Vaughan, 'Charles Booth and the Mapping of Poverty', in *Mapping Society: The Spatial Dimensions of Social Cartography* (London: UCL Press, 2018), pp. 61–92.

40 *Manchester Guardian*, 17 April 1889, cited in A. Kershen, 'Henry Mayhew and Charles Booth: Men of Their Time', in G. Alderman and C. Holmes (eds), *Outsiders & Outcasts: Essays in Honour of William J. Fishman* (London: Duckworth, 1993), p. 113.

41 Charles Booth, 'The Inhabitants of Tower Hamlets (School Board Division), Their Condition and Occupations', *Journal of the Royal Statistical Society*, 50(2), June 1887, p. 376.

42 Francis Galton, 'The Charms of Statistics', in *Natural Inheritance* (New York: Macmillan, 1889), p. 62.

43 Quoted in Daniel J. Kevles, *In the Name of Eugenics: Genetics and the Uses of Human Heredity* (Berkeley: University of California Press, 1985), p. 7.

44 Ruth Schwartz Cowan, 'Francis Galton's Statistical Ideas: The Influence of Eugenics', *Isis*, 63(4), December 1972, p. 510.

45 Cowan, p. 510.

46 Francis Galton, *Memories of My Life* (London: Methuen & Co., 1908), pp. 315–16.

47 Francis Galton, *Narrative of an Explorer in Tropical South Africa* (Minerva Library of Famous Books, 1853 (1889)), pp. 53–4.

48 Francis Galton, 'Statistical Inquiries into the Efficacy of Prayer', *Fortnightly Review*, 12, 1872, pp. 125–35.

49 As quoted in Porter, *The Rise of Statistical Thinking*, p. 133.

50 Francis Galton, *Hereditary Genius: An Inquiry into Its Laws and Consequences*, first published 1869, second edn 1892; third corrected proof of the first electronic edn, 2000, p. 1. http://galton.org.

51 Galton, *Hereditary Genius*, p. 1.

52 Francis Galton, 'Hereditary Character and Talent', *Macmillan's Magazine*, 12, 1865, pp. 157–66. http://galton.org.

53 Galton, *Hereditary Genius*, p. 341.

54 Galton, *Hereditary Genius*, pp. 338–9.

55 See Stigler, pp. 267–8; Galton quote is from *Hereditary Genius*, p. 29.

56 Galton, *Hereditary Genius*, p. 14.

57 As quoted in Karl Pearson, *The Life, Letters and Labours of Francis Galton, Vol. 3, Part A: Correlation, Personal Identification and Eugenics* (Cambridge: Cambridge University Press, 1930 (2011)), p. 30.

58 Quoted in Kevles, p. 14.

59 Quoted in Martin Brookes, *Extreme Measures: The Dark Visions and Bright Ideas of Francis Galton* (London: Bloomsbury, 2004), p. 238.

60 Stephen M. Stigler, *Statistics on the Table: The History of Statistical Concepts and Methods* (Cambridge, MA: Harvard University Press, 1999), p. 6.

61 Pearson, p. 57.

62 Galton, *Natural Inheritance*, p. 62.

63 See Adam Cohen, *Imbeciles: The Supreme Court, American Eugenics, and the Sterilization of Carrie Buck* (New York: Penguin, 2016).

64 Buck *vs* Bell, 274 US 200 (1927).

65 William E. Leuchtenburg, *The Supreme Court Reborn: The Constitutional Revolution in the Age of Roosevelt* (New York: Oxford University Press, 1995), p. 15.

66 Alexandra Minna Stern, 'Sterilized in the Name of Public Health: Race, Immigration, and Reproductive Control in Modern California', *American Journal of Public Health*, 95(7), 2005, pp. 1128–38.

67 E. B. Boudreau, '"Yea, I have a Goodly Heritage": Health Versus Heredity in the Fitter Family Contests, 1920–1928', *Journal of Family History*, 30(4), 2005, pp. 366–87.

68 C. B. Davenport, 'Report of Committee on Eugenics', *Journal of Heredity*, 1(2), 1910, pp. 126–9.

69 Kevles, p. 79.

70 Quoted in Stephen Jay Gould, *The Mismeasure of Man* (New York and London: W. W. Norton & Co., revised and expanded edn, 1996 (1981)), p. 181.

71 Quoted in Gould, p. 184.

72 Quoted in Leila Zenderland, *Measuring Minds: Henry Herbert Goddard and the Origins of American Intelligence Testing* (Cambridge: Cambridge University Press, 1998 (2001)), p. 301.

73 Zenderland, p. 175.

74 Henry H. Goddard, 'Mental Tests and the Immigrant', *The Journal of Delinquency*, vol. II, no. 5, September 1917, p. 251.

75 Edward A. Steiner, 'The Fellowship of the Steerage', in *On the Trail of the Immigrant* (New York: Fleming H. Revell Company, 1906), p. 35.

76 Quoted in Gould, p. 226.

77 Quoted in Gould, p. 253.

78 See Desrosières, p. 326.

79 P. Lazarsfeld, 'Notes sur l'histoire de la quantification en sociologie: les sources, les tendances, les grands problèmes', in *Philosophie des sciences sociales* (Paris: Gallimard, 1970), pp. 317–53. Quoted in Desrosières, p. 20.

80 Alfred Binet, *Modern Ideas About Children* (Menlo Park, CA: Susan Heisler, 1909 (1975)), pp. 106–7. Quoted in Robert J. Sternberg (ed.), *The Cambridge Handbook of Intelligence*, second edn (Cambridge: Cambridge University Press, 2020), p. 1,062.

81 Shilpa Jindia, 'Belly of the Beast: California's Dark History of Forced Sterilizations', *The Guardian*, 30 June 2020. https://www.theguardian.com/us-news/2020/jun/30/california-prisons-forced-sterilizations-belly-beast.

82 Associated Press, 'China Cuts Uighur Births with IUDs, Abortion, Sterilization', 29 June 2020. https://apnews.com/article/ap-top-news-international-news-weekend-reads-china-health-269b3de1af34e17c1941a5 14f78d764c.

83 David J. Smith, 'Biological Determinism and the Concept of Mental Retardation: The Lesson of Carrie Buck', paper presented at the Annual Convention of the Council for Exceptional Children, April 1993.

CHAPTER 8: THE BATTLE OF THE STANDARDS

1   De Simone and Treat, NIST, p. 90.

2   '"Metric Martyr" Convicted', *The Guardian*, 9 April 2001. https://www.theguardian.com/uk/2001/apr/09/2.

3   British Weights and Measures Association, 'Metric Culprits'. http://bwma.org.uk/wp-content/uploads/2019/03/Metric-Culprits-.pdf.

4   Richard Savill, 'Protest as the Metric "Martyrs" Face Court', *The Telegraph*, 14 June 2001. https://www.telegraph.co.uk/news/uknews/1309043/Protest-as-the-metric-martyrs-face-court.html.

5   Fergus Hewison, 'General Election 2019: Did a Bunch of Bananas Lead to Brexit?' BBC News, 20 November 2019. https://www.bbc.co.uk/news/election-2019-50473654.

6   Joshua Rozenberg, 'Metric Martyrs Lose Their Fight', *The Telegraph*, 19 February 2002. https://www.telegraph.co.uk/news/uknews/1385303/Metric-martyrs-lose-their-fight.html.

7   Hewison, BBC News, 20 November 2019.

8   'EU Gives Up on "Metric Britain"', BBC News, 11 September 2007. http://news.bbc.co.uk/1/hi/uk/6988521.stm.

9   'From George Washington to the United States Senate and House of Representatives, 8 January 1790', Founders Online, National Archives. https://founders.archives.gov/documents/Washington/05-04-02-0361. Original source: Dorothy Twohig (ed.), *The Papers of George Washington, Presidential Series*, vol. 4, 8 September 1789–15 January 1790 (Charlottesville: University Press of Virginia, 1993), pp. 543–9.

10 See, e.g., John J. McCusker, 'Weights and Measures in the Colonial Sugar

Trade: The Gallon and the Pound and Their International Equivalents',
The William and Mary Quarterly, 30(4), October 1973.

11 'Thomas Jefferson, Itinerary, Monticello to Washington, D.C., with
Distances', 30 September 1807. Manuscript/mixed material. Retrieved
from the Library of Congress. www.loc.gov/item/mtjbib017726/.

12 'Thomas Jefferson, Autobiography Draft Fragment, January 6 through
July 27'. 27 July 1821. Manuscript/mixed material. Retrieved from the
Library of Congress. www.loc.gov/item/mtjbib024000/.

13 'Thomas Jefferson to James Clarke, 5 September 1820'. Manuscript/mixed
material. Retrieved from the Library of Congress. www.loc.gov/item/
mtjbib023884/.

14 Keith Martin, 'Pirates of the Caribbean (Metric Edition)', National
Institute of Standards and Technology, 19 September 2017. https://www.
nist.gov/blogs/taking-measure/pirates-caribbean-metric-edition.

15 Jefferson to William Short, 28 July 1791, quoted by C. Doris Hellmann,
'Jefferson's Efforts Towards Decimalization of United States Weights and
Measures', *Isis*, 16 (1931), p. 286.

16 Daniel V. De Simone and Charles F. Treat, 'A History of the Metric
System Controversy in the United States', NIST, Special Publication 345-
10. United States Department of Commerce, August 1971, p. 18.

17 Charles Davies, *The Metric System, Considered with Reference to Its
Introduction into the United States; Embracing the Reports of the Hon.
John Quincy Adams, and the Lecture of Sir John Herschel* (New York and
Chicago: A. S. Barnes and Co., 1871), pp. 267–8.

18 Kula, p. 267.

19 Alder, pp. 330–1.

20 Maria Teresa Borgato, 'The First Applications of the Metric System in
Italy', *The Global and the Local: The History of Science and the Cultural
Integration of Europe*, M. Kokowski (ed.), Proceedings of the 2nd ICESHS
(Cracow, Poland, 6–9 September 2006).

21 Edward Franklin Cox, 'The Metric System: A Quarter-Century of
Acceptance (1851–1876)', *Osiris*, 13, 1958, p. 363.

22 Charles L. Killinger, *The History of Italy* (Connecticut and London:
Greenwood Press, 2002), p. 1.

23 Vanessa Lincoln Lambert, 'The Dynamics of Transnational Activism: The
International Peace Congresses, 1843–51', *The International History Review*,
38(1), 2016, pp. 126–47.

24 Report of the Proceedings of the . . . General Peace Congress (London:
Charles Gilpin, 1849), p. 12.

25 Quoted in Richard R. John, 'Projecting Power Overseas: U.S. Postal

Policy and International Standard-Setting at the 1863 Paris Postal Conference', *Journal of Policy History*, 27(3), 2015, p. 430.

26 Zupko, p. 238.

27 Charles Sumner, 'The Metric System of Weights and Measures', speech in the Senate of the United States, 27 July 1866.

28 Cox, p. 377.

29 Simon Schaffer, 'Metrology, Metrication, and Values', in Bernard Lightman (ed.), *Victorian Science in Context* (Chicago and London: University of Chicago Press, 1997), p. 442.

30 Crease, p. 156.

31 See Tessa Morrison, 'The Body, the Temple, and the Newtonian Man Conundrum', *Nexus Network Journal*, 12 (2), 2010, pp. 343–52; and original Newton source: http://www.newtonproject.ox.ac.uk/view/texts/normalized/THEM00276.

32 Charles Piazzi Smyth, *Our Inheritance in the Great Pyramid* (New York: Cambridge University Press, 1864 (2012)), p. 372.

33 Charles Piazzi Smyth, *Our Inheritance in the Great Pyramid* (fourth edn, 1880); published as *The Great Pyramid: Its Secrets and Mysteries Revealed* (New York: Gramercy Books, 1978), pp. 546–8.

34 Piazzi Smyth, *Our Inheritance in the Great Pyramid*, p. 10.

35 Quoted in Schaffer, p. 451.

36 Charles Piazzi Smyth, *Life and Work at the Great Pyramid*, vol. 1 (Edinburgh: Thomas Constable, 1867), p. 299.

37 Charles A. L. Totten, *An Important Question in Metrology Based Upon Recent and Original Discoveries: A Challenge to "The Metric System," and an Earnest Word with the English-Speaking Peoples on Their Ancient Weights and Measures* (New York: John Wiley & Sons, 1884), pp. xi–xii.

38 Totten, p. xiv.

39 Totten, p. 2.

40 Totten, p. 45.

41 Quoted in De Simone and Treat, NIST, p. 75.

42 Piazzi Smyth, *Our Inheritance in the Great Pyramid*, p. 182.

43 De Simone and Treat, NIST, pp. 75–8.

44 George Orwell, *1984* (1949, copyright renewed 1977) (New York: Houghton Mifflin Harcourt Publishing Co., 1983), p. 83.

45 '"Imperial Vigilante" Wins Legal Appeal', BBC News, 31 October 2002. http://news.bbc.co.uk/1/hi/england/2384065.stm.

46 Totten, p. x.

47 Benjamin Kentish, 'Half of Leave Voters Want to Bring Back the Death Penalty After Brexit', *The Independent*, 30 March 2017. https://www.

independent.co.uk/news/uk/politics/brexit-poll-leave-voters-death-penalty-yougov-results-light-bulbs-a7656791.html.

48 National Academies of Sciences, Engineering, and Medicine; Division on Earth and Life Studies; Nuclear and Radiation Studies Board, *Adopting the International System of Units for Radiation Measurements in the United States: Proceedings of a Workshop* (Washington, DC: National Academies Press, 2017), 1, Introduction and Context. Available from https://www.ncbi.nlm.nih.gov/books/NBK425560/.

49 Both quoted in John Bemelmans Marciano, *Whatever Happened to the Metric System? How America Kept Its Feet* (New York: Bloomsbury USA, 2014 (2015)), p. 244.

50 Stewart Brand, 'Stopping Metric Madness!' *New Scientist*, 88, 30 October 1980, p. 315.

51 Paul L. Montgomery, '800, Putting Best Foot Forward, Attend a Gala Against Metrics', *The New York Times*, 1 June 1981.

52 A. V. Astin and H. Arnold Karo, National Bureau of Standards, 'Refinement of Values for the Yard and the Pound', 30 June 1959. https://www.ngs.noaa.gov/PUBS_LIB/FedRegister/FRdoc59-5442.pdf.

53 Shane Croucher, 'Video: Fox Host Tucker Carlson Attacks "Inelegant, Creepy" Metric System that the U.S. Alone Has Resisted', *Newsweek*, 6 June 2019.

CHAPTER 9: FOR ALL TIMES, FOR ALL PEOPLE

1 Laurence Stern, *The Life and Opinions of Tristram Shandy, Gentleman and a Sentimental Journey Through France and Italy (Volume 1)* (London: Macmillan and Co. Ltd, 1900 (1759)), book II, chapter XIX, p. 131. Full quote: 'The laws of nature will defend themselves – but error – (he would add, looking earnestly at my mother) – error, Sir, creeps in thro' the minute holes and small crevices which human nature leaves unguarded.'

2 Report from Borda, Lagrange, Laplace, Monge, and Condorcet, 'On the Choice of a Unit of Measurement', presented to the Académie Royale des Sciences, 19 March 1791.

3 G. Girard, 'The Washing and Cleaning of Kilogram Prototypes at the BIPM', Bureau International des Poids et Mesures (BIPM), 1990.

4 Terry Quinn, *From Artefacts to Atoms: The BIPM and the Search for Ultimate Measurement Standards* (New York: Oxford University Press, 2012), pp. 341–6.

5 Quinn, p. 365.

6 James Vincent, 'The Kilogram Is Dead; Long Live the Kilogram', *The Verge*, 13 November 2018. https://www.theverge.com/2018/11/13/

18087002/kilogram-new-definition-kg-metric-unit-ipk-measurement.

7 'American Aristotle' and 'without precedent in the history of the Earth': https://aeon.co/essays/charles-sanders-peirce-was-americas-greatest-thinker. 'Most original and most versatile intellect': Max H. Fisch, 'Introductory Note', in Sebeok, *The Play of Musement*, p. 17, quoted in Joseph Brent, *Charles Sanders Peirce: A Life, Revised and Enlarged Edition* (Bloomington: Indiana University Press, 1998), p. 2.

8 Brent, p. 41

9 William James, *Essays in Philosophy, 1876–1910* (Cambridge, MA: Harvard University Press; 1978); 'The Pragmatic Method', p. 124.

10 Joseph Brent, 'Studies in Meaning', Charles S. Peirce Papers, p. 15; quoted in Louis Menand's *The Metaphysical Club* (London: Flamingo, HarperCollins Publishers imprint, 2001), p. 160.

11 Brent, *Charles Sanders Peirce: A Life*, pp. 48–9.

12 Brent, *Charles Sanders Peirce: A Life*, p. 49.

13 Charles Sanders Peirce, 'A Guess at the Riddle', 1887; quoted in Brent, *Charles Sanders Peirce: A Life*, p. 1.

14 Schaffer, p. 440.

15 F. K. Richtmyer, 'The Romance of the Next Decimal Place', *Science*, 1 January 1932, 75 (1931), pp. 1–5.

16 Ronald Robinson and John Gallagher with Alice Denny, *Africa and the Victorians: The Official Mind of Imperialism*, second edn (London: Macmillan, 1961 (1981)), p. 12.

17 Both quoted from Daniel Headrick, 'A Double-Edged Sword: Communications and Imperial Control in British India', *Historical Social Research*, 35(1), 2010, p. 53.

18 Zupko, p. 208 (archive.org copy).

19 Caroline Shenton, *The Day Parliament Burned Down* (Oxford: Oxford University Press, 2012), p. 212.

20 Quoted in Quinn, 2012, p. xxviii.

21 Quoted in Schaffer, p. 444.

22 Quoted in Schaffer, p. 445.

23 Schaffer, p. 448.

24 Victor F. Lenzen, 'The Contributions of Charles S. Peirce to Metrology', *Proceedings of the American Philosophical Society*, 109(1), 18 February 1965, pp. 29–46. Quoth Peirce: 'The ratio of the meter to the yard is still a matter of considerable uncertainty', *Report of the Superintendent*, 30 June 1884, p. 81.

25 Quoted in Robert P. Crease, 'Charles Sanders Peirce and the First Absolute Measurement Standard', *Physics Today*, vol. 62, no. 12, 2009, p. 39.

26 Ralph Barton Perry, *The Thought and Character of William James* (Boston: Atlantic/Little, Brown, 1935), I, p. 536.

27 Crease, *World in the Balance*, p. 198.

28 'Annual Report of the Director, United States Coast and Geodetic Survey, to the Secretary of Commerce', U.S. Coast and Geodetic Survey, 1881, p. 28.

29 Charles Sanders Peirce, 'The Fixation of Belief', *Popular Science Monthly*, 12 (November 1877).

30 Peirce, 'The Fixation of Belief'.

31 Charles Sanders Peirce, 'Philosophy and the Sciences: A Classification', from Justus Buchler (ed.), *Philosophical Writings of Peirce* (New York: Dover Publications, 1955), p. 73.

32 From the Lowell Lectures in 1903, quoted in Cornelis de Waal, *Peirce: A Guide for the Perplexed* (London and New York: Bloomsbury, 2013), p. 113.

33 1897 [*c.*]; Notes on Religious and Scientific Infallibilism [R]; CP 1.13–14; from the Robin Catalogue: A. MS., G-c.1897-2, 4 pp. and 7 pp. http://www.commens.org/dictionary/term/fallibilism/page.

34 1893 [*c.*]; Fallibilism, Continuity, and Evolution [R]; CP 1.147-149. Manuscript; from the Robin Catalogue: A. MS., G-c.1897-5, 57 pp. http://www.commens.org/dictionary/term/fallibilism/page.

35 E.g. Newton's mechanics; Huygen's optics; Maxwell's electrodynamics.

36 From Michelson's dedication of the Ryerson Physical Laboratory at the University of Chicago. Quoted in *The University of Chicago Annual Register, 1 July 1894 to 1 July 1895* (Chicago: University of Chicago Press, 1895), p. 81.

37 Malcolm W. Browne, 'In Centennial of One of Its Biggest Failures, Science Rejoices', *The New York Times*, 28 April 1987. https://www.nytimes.com/1987/04/28/science/in-centennial-of-one-of-its-biggest-failures-science-rejoices.html, accessed 25 January 2021.

38 Albert A. Michelson and Edward W. Morley, 'On a Method of Making the Wave-length of Sodium Light the Actual and Practical Standard of Length', *Philosophical Magazine*, series 5 (1876–1900), 24(151), 1887.

39 Harvey T. Dearden, 'How Long Is a Metre?' *Measurement and Control*, vol. 47(1), 2014, pp. 26–7. © The Institute of Measurement and Control 2014.

40 Albrecht Fölsing, *Albert Einstein: A Biography* (London: Penguin Group, 1998), p. 219.

41 Quinn, 2012, p. XXVI.

42 From here: https://www.youtube.com/watch?v=bjVfL8uNkUk&ab_channel=ArvinAsh.

43 Much of this explanation comes from https://www.nist.gov/si-redefinition/kilogram/kilogram-mass-and-plancks-constant and https://www.nist.gov/si-redefinition/kilogram/kilogram-kibble-balance. A joule-second, or J s, is a unit of energy × time, with 1 J s approximately equal to lifting an apple a metre in the air.

44 Bruno Latour, *Science in Action* (Cambridge, MA: Harvard University Press, 1987), pp. 2–4.

45 Matthew 7:16 (KJV).

CHAPTER 10: THE MANAGED LIFE

1 Hartmut Rosa, *The Uncontrollability of the World*, James C. Wagner (trans.) (Cambridge: Polity Press, 2020), p. 2. Originally published in German as *Unverfügbarkeit* © 2018 Residenz Verlag GmbH, Salzburg-Wein.

2 Kenneth G. W. Inn, 'Making Radioactive Lung and Liver Samples for Better Human Health', National Institute of Standards and Technology, 29 October 2019. https://www.nist.gov/blogs/taking-measure/making-radioactive-lung-and-liver-samples-better-human-health.

3 Marty Ahrens, 'Home Fires Started by Smoking', National Fire Protection Association, January 2019. https://www.nfpa.org/News-and-Research/Data-research-and-tools/US-Fire-Problem/Smoking-Materials.

4 Henry Fountain, 'You Get What You Pay for: Peanut Butter with a Pedigree', *The New York Times*, 8 June 2003. Note: the quote is only available in the digitized version of the physical article, not in the web version. Many thanks to John Farrier for help in locating this.

5 Allison Loconto and Lawrence Busch, 'Standards, Techno-Economic Networks, and Playing Fields: Performing the Global Market Economy', *Review of International Political Economy*, 17 (3), 2010, pp. 507–36.

6 Mark Aldrich, 'Public Relations and Technology: The "Standard Railroad of the World" and the Crisis in Railroad Safety, 1897–1916', *Pennsylvania History: A Journal of Mid-Atlantic Studies*, 74(1), Winter 2007, p. 78.

7 'Mainline passenger rail service had a considerably worse fatality rate at 0.43 per billion passenger miles. This is about four times the risk for bus passengers and six times the risk in commercial aviation.' Ian Savage, 'Comparing the Fatality Risks in United States Transportation Across Modes and Over Time', *Research in Transportation Economics*, 43(1), July 2013, pp. 9–22.

8 It's worth noting that there are plenty of precedents to Muller's observations. See, for example, V. F. Ridgway, 'Dysfunctional

Consequences of Performance Measurements', *Administrative Science Quarterly*, 1(2), September 1956, pp. 240–7. Or Peter Drucker's 1954 book *The Practice of Management*, from where comes the adage 'what gets measured gets managed'. Similar criticisms can be found in the work of W. Edwards Deming, Henry Mintzberg, etc. See https://www.theguardian.com/business/2008/feb/10/businesscomment1. And Goodhart's law – 'Any observed statistical regularity will tend to collapse once pressure is placed upon it for control purposes' – then turned into 'When a measure becomes a target, it ceases to be a good measure' in a paper published in 1997 by anthropologist Marilyn Strathern.

9 Jerry Z. Muller, *The Tyranny of Metrics* (Princeton: Princeton University Press, 2018), pp. 3–4.

10 Alfred D. Chandler, Jr., 'The Emergence of Managerial Capitalism', *The Business History Review*, 58(4), Winter 1984, p. 473. See also Chandler, *The Visible Hand: The Managerial Revolution in American Business*.

11 Nathan Rosenberg (ed.), 'Special Report of Joseph Whitworth', from *The American System of Manufactures*, p. 387; quoted in David A. Hounshell, *From the American System to Mass Production, 1800–1932* (Baltimore and London: Johns Hopkins University Press, 1984), p. 61.

12 Frederick W. Taylor, 'Testimony to the House of Representatives Committee', quoted in Rakesh Khurana, *From Higher Aims to Hired Hands: The Social Transformation of American Business Schools and the Unfulfilled Promise of Management as a Profession* (Princeton: Princeton University Press, 2007), p. 93.

13 Frank B. Gilbreth, Jr. and Ernestine Gilbreth Carey, *Cheaper by the Dozen* (New York: Thomas Y. Cromwell Co., 1948), p. 127.

14 'I don't fault Truman for dropping the nuclear bomb. The US–Japanese War was one of the most brutal wars in all of human history – kamikaze pilots, suicide, unbelievable. What one can criticize is that the human race prior to that time – and today – has not really grappled with what are, I'll call it, "the rules of war." Was there a rule then that said you shouldn't bomb, shouldn't kill, shouldn't burn to death 100,000 civilians in one night?

'LeMay said, "If we'd lost the war, we'd all have been prosecuted as war criminals." And I think he's right. He, and I'd say I, were behaving as war criminals. LeMay recognized that what he was doing would be thought immoral if his side had lost. But what makes it immoral if you lose and not immoral if you win?'

From the documentary *The Fog of War*. Quoted here: 'Robert McNamara', Michael Tomasky, 6 July 2009. https://www.

theguardian.com/commentisfree/michaeltomasky/2009/jul/06/
robert-mcnamara-vietnam.

15 Edward N. Luttwak, *The Pentagon and the Art of War; The Question of
Military Reform* (New York: Simon & Schuster, 1985 (1986)), p. 268.

16 Quoted in Nick Turse, *Kill Anything That Moves: The Real American War
in Vietnam* (New York: Metropolitan Books, 2013), pp. 47–8.

17 Turse, p. 49.

18 Gary Wolf, 'The Data-Driven Life', *The New York Times*, 28 April
2010. https://www.nytimes.com/2010/05/02/magazine/02self-
measurement-t.html.

19 April Dembosky, 'Invasion of the Body Hackers', *The Financial Times*, 10 June
2011. https://www.ft.com/content/3ccb11a0-923b-11e0-9e00-00144feab49a.

20 Evgeny Morozov, *To Save Everything Click Here: Technology, Solutionism,
and the Urge to Fix Problems That Don't Exist* (New York: Penguin, 2013),
p. 223.

21 Wolf, 'The Data-Driven Life'.

22 Zach Church, 'Google's Schmidt: "Global Mind" Offers New
Opportunities', *MIT News*, 15 November 2011. https://news.mit.
edu/2011/schmidt-event-1115.

23 Shoshana Zuboff, *The Age of Surveillance Capitalism* (New York: Public
Affairs, 2019), pp. 7–8.

24 Drew Harwell, 'Wanted: The "Perfect Babysitter" Must Pass AI
Scan for Respect and Attitude', *The Washington Post*, 23 November
2018. https://www.washingtonpost.com/technology/2018/11/16/
wanted-perfect-babysitter-must-pass-ai-scan-respect-attitude/.

25 Christopher Ingraham, 'An Insurance Company Wants You to Hand
Over Your Fitbit Data So It Can Make More Money. Should You?' *The
Washington Post*, 25 September 2018. https://www.washingtonpost.com/
business/2018/09/25/an-insurance-company-wants-you-hand-over-your-
fitbit-data-so-they-can-make-more-money-should-you/.

26 Drew Harwell, 'A Face-Scanning Algorithm Increasingly Decides
Whether You Deserve the Job', *The Washington Post*, 6 November 2019.
https://www.washingtonpost.com/technology/2019/10/22/ai-hiring-
face-scanning-algorithm-increasingly-decides-whether-you-deserve-
job/.

27 Jeffrey Dastin, 'Amazon Scraps Secret AI Recruiting Tool That Showed
Bias Against Women', Reuters, 11 October 2018. https://www.reuters.com/
article/us-amazon-com-jobs-automation-insight-idUSKCN1MK08G.

28 Adam D. I. Kramer, Jamie E. Guillory and Jeffrey T. Hancock,
'Experimental Evidence of Massive-Scale Emotional Contagion Through

Social Networks', *Proceedings of the National Academy of Sciences*, 111(24), June 2014.

29 Joanna Kavenna, 'Shoshana Zuboff: "Surveillance Capitalism Is an Assault on Human Autonomy"', *The Guardian*, 4 October 2019. https://www.theguardian.com/books/2019/oct/04/shoshana-zuboff-surveillance-capitalism-assault-human-automomy-digital-privacy.

30 Fyodor Dostoevsky, *Notes from Underground*, translated from the Russian by Richard Pevear and Larissa Volokhonsky (New York: Alfred A. Knopf, 2004), p. 23.

31 Lily Kuo, 'China Bans 23m from Buying Travel Tickets as Part of "Social Credit" System', *The Guardian*, 1 March 2019. https://www.theguardian.com/world/2019/mar/01/china-bans-23m-discredited-citizens-from-buying-travel-tickets-social-credit-system.

32 As quoted in Simina Mistreanu, 'Life Inside China's Social Credit Laboratory', *Foreign Policy*, 3 April 2018. https://foreignpolicy.com/2018/04/03/life-inside-chinas-social-credit-laboratory/.

33 Genia Kostka, 'China's Social Credit Systems and Public Opinion: Explaining High Levels of Approval' (23 July 2018). Available at SSRN: https://ssrn.com/abstract=3215138 or http://dx.doi.org/10.2139/ssrn.3215138.

34 Porter, *Trust in Numbers*, 1995, p. ix.

35 Amanda Mull, 'What 10,000 Steps Will Really Get You', *The Atlantic*, 31 May 2019. https://www.theatlantic.com/health/archive/2019/05/10000-steps-rule/590785/. See also I-Min Lee, Eric J. Shiroma, Masamitsu Kamada, et al., 'Association of Step Volume and Intensity with All-Cause Mortality in Older Women', https://jamanetwork.com/journals/jamainternalmedicine/fullarticle/2734709.

36 Catrine Tudor-Locke and David R. Bassett Jr., 'How Many Steps/Day Are Enough? Preliminary Pedometer Indices for Public Health', *Sports Medicine*, 34(1), 2004, pp. 1–8.

37 I-Min Lee et al., 'Association of Step Volume and Intensity with All-Cause Mortality in Older Women', *JAMA Internal Medicine*, 179(8), 2019, pp. 1,105–12.

38 K. T. Hallam, S. Bilsborough, and M. de Courten, '"Happy Feet": Evaluating the Benefits of a 100-Day 10,000 Step Challenge on Mental Health and Wellbeing', *BMC Psychiatry*, 18(1), 24 January 2018. doi: 10.1186/s12888-018-1609-y. PMID: 29361921; PMCID: PMC5781328.

39 https://twitter.com/chr1sa/status/721198400150966274?lang=en-gb.

40 Dembosky, 'Invasion of the Body Hackers', *The Financial Times*.

41 Elizabeth Lopatto, 'Jeff Bezos Appreciates Your Efforts to Get Jeff

Bezos to Space', *The Verge*, 20 July 2021. https://www.theverge.com/2021/7/20/22585470/jeff-bezos-blue-origin-space-amazon-customers.

42 Lauren Kaori Gurley, 'Amazon Denied a Worker Pregnancy Accommodations. Then She Miscarried', *Motherboard for Vice*, 20 July 2021. https://www.vice.com/en/article/g5g8eq/amazon-denied-a-worker-pregnancy-accommodations-then-she-miscarried.

43 Rosa, pp. 6–7.

44 Rosa, p. 7.

45 Rosa, p. 60.

LIST OF PERMISSIONS

TEXT PERMISSIONS

INDEX

Page numbers in *italics* relate to illustrations.